Are Aliens Real? Aliens and UFOs Proof

By Sean Keyhoe

***

Published by Kindle

Are Aliens Real? Aliens and UFOs Proof

By Sean Keyhoe

Table of Contents

Preface

I think maybe I hate UFOs...actually I hate the stigma that is attached to UFO sightings and I think I know why. UFO sightings always seem to be grouped together into a single paranormal basket that includes crop circles, a moon landing hoax and a million and one other conspiracies. It seems you cannot have one without the other.

But seriously, I have held a lifelong interest in UFOs and aliens since my own personal experience many years ago. Coming home late at night after having gone to see a movie with friends, I caught the last running bus that left the main terminal from downtown at 11:30 PM. It dropped me about a mile from home; as I made my way up the hill towards our house it must have been well after midnight.

The street was forested on both sides and above the tree line I saw three oval shaped lights; they moved alongside me almost as if they were following my pace.

They looked to be on the horizon about as far as I would expect clouds to be. One of the three broke off from the others and zipped away into the night. The other two continued along for some seconds before also accelerating and flying off together. If I could have snapped a photo at the time, it would have looked something like this:

At the time I felt a twinge of fear and excitement and soon thereafter felt bad that I was alone and did not have the opportunity to capture the sight to share with others. I have thought about that experience often and wondered what I did actually see that night. The time and date of when this happened is no longer important to me. To this day I cannot fully explain what it may have been.

Throughout the years I have kept an active interest in UFOs and Alien activity and it seems that more and more compelling evidence emerges on an almost daily basis that might somehow reveal genuine proof of the existence of Aliens and UFOs among us. In the late 1980's I was lucky enough to meet an individual who claimed to be an "insider" concerning matters of UFO and Aliens. I shared with him the experience I had as a teenager and he shared with me details of his life and work as a high level government "employee". He had a vast knowledge of details that I can only briefly mention at this time. The most important aspect of our relationship, and the underlying reason behind the publishing of this book, is the fact that he entrusted to me documentation and evidence that he felt was overwhelming evidence for the existence of extraterrestrials. This is the first major purpose of this book, that is, to share with you the reader, the details that have been passed on to me from this individual.

The second major purpose of this book is to save you time! This book presents a collection and a distillation of popular evidence; some of which has already been partially documented in the past but also some of which has never been revealed until now. My goal in presenting this collection of work is to provide the most compelling evidence from firsthand accounts and to provide the most accurate transcriptions from dependable audio and video sources.

I must explain that some of the testimonials and commentary that I have purposely included in this book, appear to me as delusions, and I give reasons why I feel they are in the pages to follow. The reason I have included them is for several reasons. The first is because those testimonials are widespread and well known, that is, these individuals have been presented as believable, and I feel it is important to provide a dissenting opinion if no one else has *and* if that is honestly how I feel. This is especially important since these viewpoints are often the first ones found when someone first begins to actively look into the topic of Aliens and UFOs.

That is, when one begins to search for information on Aliens and UFOs, the individuals from Section 1 of this book, are usually the first ones that appear in the popular literature and online community.

The reason this book can save you time is due to the huge volume of false, irrelevant and misleading information that has flooded the topic of UFOs and Aliens. I have spent hundreds of hours reading and watching video on this subject over the past years and I can tell you that the area of UFO research can become a black hole of never ending information that splits off into hundreds of tangents. This book is a summary of the essence of what I have compiled and found most relevant, compelling and valuable on the subject of UFOs and Aliens to date. Therefore my hope is that this will be useful to you as a primer (for those who are new to UFO research), as a reference (for those who are curious) or as an addition to your existing library (for those who are experienced UFO researchers).

Thank you for choosing this book! Please share, like and send a gift copy to a friend!

~ Sean Keyhoe

Introduction

The European Union Times Online Newspaper published a report on January 22nd, 2013. They were relating coverage of the upcoming World Economic Forum which took place in Switzerland from January 23rd to January 27th

Dmitry Medvedev

In this report they quoted the Russian President Dmitry Medvedev as saying the following:

"Along with the briefcase with nuclear codes, the president of the country is given a special 'top secret' folder. This folder in its entirety contains information

about aliens who visited our planet… Along with this, you are given a report of the absolutely secret special service that exercises control over aliens on the territory of our country… More detailed information on this topic you can get from a well-known movie … I will not tell you how many of them are among us because it may cause panic."

Western journalists were startled at this comment to say the least. In fact the original Sky News television coverage quickly cut away from the above interview coverage mid-sentence. Some clips only show about half of the above statement. Some people quoted the Russian President as referring to the movie 'Men-In-Black' but I believe he was referring to the Russian movie of a similar name.

It is interesting to note that western media has typically defined 'men-in-black' as being government agents that work behind the scenes carrying out top secret military related tasks. This view has been perpetuated by the comedy movies of the same name starring Will Smith and Tommy Lee Jones.

In Russia, on the other hand, 'men-in-black' are commonly known as actual alien beings that disguise themselves and work in cooperation with government departments and high level insiders.

The Russian movie that was referred to by President Medvedev is actually titled 'Hidden Territories' as translated from Russian. The link to the movie from the EU Times news report has since become inactive. Luckily I was able to access the movie and the full transcript is presented below.

Hidden Territories Transcript:

In the center of the mountainous area of Carpathians in the Western Ukraine in 1946, a child was found. Homeless and looked like the age of seven. He could not (give) his name or his home address (likely) as a consequence of the war. At the orphanage odd things about this boy were discovered. The children reported that during his sleep his body elevated over the bed. He also spoke with trees and birds and could resurrect a dead animal. This amazing boy could also predict the future. He predicted that 16 years later, the first man in orbit would be a Russian named "Yuri". After a while people in Police uniforms came for the child. They told the family he was staying with that they had found his parents just so they could take him away.

Afterwards he was visited by a man from the security department and questioned by other officials far away at Lubyanka which is the location of KGB headquarters.

The secret about "Men in Black".

Anton Anfalov is also known as "Crimean Malodarom". Like a movie hero, he seeks the truth about flying saucers. Anton's interest in this area increased when he saw an object with a complex shape over the city. The object was huge and black with shining lights around.

"I usually know all airplanes, helicopters and I can distinguish earth aircraft from those that are not; and this was a totally shocking site. It had a red flashing light in front, and yellow outlined contours around. Its length was not more than 30 to 50 meters."

The object passed silently over Anton Anfalov and hid into the distance. Since then, this University professor set a goal to find out who could operate such an aircraft. The research took him far away and he discovered that there is a worldwide alien conspiracy.

Military test pilot and a hero of the Soviet Union, Nikolai Antoshkin, has excellent knowledge of the latest developments in the most secret aircraft models and can distinguish them from UFOs. He says that Russian airspace is not only used by man-made vessels. Here is just one example:

Nikolai Antoshkin: "Before the sun had set in one of our distant areas, a shining object appeared. It had an oval shape with some sort of orange-brown and golden colors. It seemed to us that maybe one of craft had caught fire because there were some other cases where aircraft had caught fire in the air. For some reason that was what we thought. The flight chief contacted our pilots by radio. Everybody at the scene reported that there was no fire with any of our aircraft.

Then it was decided to send a reconnaissance MIG-21; and while our MIG-21 was flying towards the object, and there were maybe two, three or four kilometers left...when suddenly the oval orange-red-brown-golden object...with huge speed flew into space at a very steep, near vertical angle. Then it turned into a spot and disappeared. Then it became clear to all our pilots that a simple machine cannot move with such a tremendous speed. The burning ellipse showed maneuvers that the finest aircrafts cannot do.

The pilots discussed for a long time what they had seen. Gradually conversations generally stopped; especially after a KGB operative assistant brought everyone together and said, 'End all talking about this, it is just a simple lightning.'

Well of course we all knew very well this was not just lightning. We have all seen lightning. And in general we know what a ball or a bolt of lightning looks like."

The amount of UFO appearances around the world have increased since World War II. This can be explained for three reasons.

The first reason is that Hitler had the opportunity to make disc shaped aircraft. Later these devices and blueprints were brought to the United States.

Secondly, the arms race helped in the creation of alternative types of aircraft and weapons.

And third, with the beginning of the nuclear age there was a danger of the destruction of our planet and it caused an increased interest from the aliens."

Alexander Semenov: "Since the amount of UFO cases began to grow, as a researcher Anton Anfalov found more evidence about the existence of UFOs on earth in the second half of the last century. Looking at some pictures from satellites of U.S. military bases he had noticed that although the photos had been modified, something wasn't hidden.

Anton Anfalov: "North of the nuclear facility Indiana springs in Nevada, there is an anomalous structure on the military firing field that could not be explained by any way because there is no existing road and it is very difficult to cross those high grounds. Information about underground bases and aliens were also confirmed by a former U.S. soldier by the name of Charles Hope."

In the 1960's Charles Hope served in the U.S. Air Force at Nellis Range which is not very far from the Indiana Springs base. He claims that he had often communicated with aliens living there. This race of aliens was called 'The Tall Whites' and their representatives live and work on earth with the approval of the U.S. government.

Anton Anfalov: "Contact with this alien race had begun a long time ago."

Researchers believe that in October 1941 in North Carolina an alien egg-shaped craft had crashed there. The crew consisted of four tall humanoids with milky white skin. Those who survived were taken to secret bunker where they died a few weeks later.

Anton Anfalov: "The report of the Americans was that they died of hunger because they could not eat the food that was served to them. They might have been clones planted specifically for us. Without tangible evidence the military generals and President Roosevelt would never have believed it. The ship and the humanoids were a diplomatic gift to the U.S. government. That was the reason to start the talks. After this event the White House had covered the unusual incident in Roswell in 1947."

A farmer William Brazel on his ranch found a disk shaped silver object and many unexplainable fragments. The military balloon 'Mobl' had crashed there which was designed for spying on the Soviet nuclear program. But the Americans had faked that and exposed fragments of an alien craft. They also took some pictures to make a documentary about it. But where there's smoke, there must be a fire. However the alien contact was already held.

Aliens met the U.S. President Dwight Eisenhower at a military base and agreed to cooperation. They offered their technology and as a return they asked to be allowed to experiment on people. Both sides agreed to do experiments with memory erasing on 2% of the U.S. population. But either the 'Tall Whites' humanoids exceeded that number or the people that came forward had increased.

Humans were also abducted by other aliens which included not only the territory of the U.S. Anastasia Erteminka is one of those unfortunate people. One evening in the center of town while she was crossing a street, she saw a bright white light and an extraordinary creature appeared.

Anastasia Erteminka: "It was a male, very tall and wearing something like a suit or alien clothes. I thought this was a toy because sometimes there are toys like this on the street." And then with a shrill metallic voice it asked Anastasia if she would like to fly into space. Because she was in a hurry to get home and prepare dinner for her child she thought that someone was kidding with her.

Anastasia Erteminka: "At that moment I looked at it and said, 'I would like that but it's very expensive'. The creature said it is not expensive and I could go whenever I decide because they are real aliens. Then I saw its eyes and they were enormous."

But after a while a beam picked her up and she found herself in a cigarette-shaped object. The put her on a chair with a screen in front of her that showed the past and the future.

Anastasia Erteminka: "I thought it was a time machine. On the one side it showed what will happen in the future. And on the other what had happened in the recent past. It was like a review of the whole life."

After this the UFO took off towards the sea. Anastasia saw the Crimea coasts and the famous 'mountain of bears'. Then a gate opened and inside the mountain there was an alien base. There she saw a huge hangar and dozens of flying saucers.

Anastasia Erteminka: "The flying saucers stood on some kind of holders and the aliens were sitting on something like stairs working or repairing something."

After this event Anastasia has only partial memory of medical experiments on other humans at the base. She was at home late that night and could not tell her friends where she was for more than eight hours. After Anastasia submitted to some hypnosis it was proved that her memory had been erased as a result of her abduction.

In the early 40's people who had seen UFOs were visited by unusual visitors. More about this is known in the U.S. but they were also seen in the Soviet Union.

Alexander Semenov: "UFO eyewitnesses or those contacted by the aliens starting from the early 1950's not only in USA but also in Europe were often visited by people dressed all in black…,black coat, black shoes, black tie and black hat. The so called 'men in black' were especially interested in witnesses who had some information about alien bases. They had just one goal, convincing people to forget about what they had seen and live without threats. But these visits caused the witnesses even greater horror. Dressed in black they looked like a human only from a distance. They had a very strange voice like a shrill sound. Sometimes cables and hoses appeared from their shoes and their skin was so pale. Also their eyes could not be seen. They also had different gadgets with themselves.

Military retiree Victor Zdorov has seen these 'men in black'. He live in an anomalous zone and has been abducted by aliens several times. The first time happened many years ago. An ordinary evening while watching T.V. with his wife; suddenly the room went dark and all sounds were gone.

Victor Zdorov: "I felt like I was going to sleep and I knew that something weird was going to happen. It looked like a levitation and we all know that levitation could not be performed by humans. So someone else was trying to lift me out of my bed. I looked up at the ceiling but there was no ceiling. All I could see was the sky and at about 100 meters above my home, there was a flying saucer."

The UFO unleashed a beam on Victor and his wife to bring them inside. He strongly confronted this and the flying saucer flew away without them. But the next time, the human curiosity prevailed. This was the way the contact began and many years later he has been forced to remain silent. Victor says he participated in many alien experiments and has visited many alien bases on earth.

Victor Zdorov: "These alien bases are here since the beginning of the Earth. Their first goal is to control the development of our planet and stop any spontaneous process that could lead to an explosion or destruction of the Earth. The second goal is to control and adjust the evolution of the humanity on Earth. And the third is to protect the humanity on Earth from any impact of various external factors like: asteroids, harmful emissions, a burst of a new star or something similar."

Victor Zdorov learned a lot about the role of extraterrestrial civilizations in the history of humanity and the world's top secrets. He wanted to share this information with his friends but they convinced him to remain silent.

Victor Zdorov: "I was often visited by men in black. They always followed me around and watched how I behaved. But if they were noticed by the neighbors then they would immediately leave. Usually they were in a group of two or three and observed everything around."

Victor doesn't remember this but this was told to him by his neighbors. He suspected that these strange visitors had erased his memory. According to Victor, he knows the role of 'men in black' from the aliens themselves.

Victor Zdorov: "Those are highly developed humanoid robots and possess a great intellect and have great potential. Also they could know any situation a few steps forward; like a clairvoyant and they can even teleport themselves. And with one look they can cause pain or erase your memory."

In the movies, we see small creatures that have come from other planets but it's possible the invented things to be mixed with the reality because almost a half century ago there were witnesses about the U.S. President meeting with aliens. "The Tall Whites"; this race is also known as Indugut.

Anton Anfalov has been in contact with his American colleagues for a long time and together they analyze these leaks of secret information. He was prepared for some shocking facts but the truth was even worse than he expected. He found that the aliens other than the military facilities were also found in the White House.

The aliens use our planet as a landing site for their interplanetary flights and use its resources. Bases are owned by "The Tall Whites" and bases exist on the moon also. Kirim Osmanov has been to one of these bases. This person had differed from the others even in his childhood. He can see things much clearer than others and he knows his future. Kirim was born in a village in middle Asia where huge orange crafts often fly above and creatures with grey suits are seen. Not so long ago he woke up with a strange feeling.

Kirim Osmanov: "I got up from bed and I went outside to the park next to our house. There was a circle-shaped space craft standing on the ground. So without being forced, but probably ordered I sat there without fear."

Kirim entered a room filled with light and sat on a chair. There was no one in the craft. The flying saucer smoothly and quietly flew in the sky. This young man looked out the window and saw our planet from the orbit.

Then the capsule with Karim inside flew over a rocky landscape and began to descend into a crater. After a long descent inside the crater, he saw some people at the bottom there.

Kirim Osmanov: "They were dressed like primitive people. Sleeveless,...actually without upper garment. But they were clean and their faces also. I can't say they were miners digging something. But I got the impression that they live there quiet and peaceful without the slavery shown on movies where people are being exploited on other planets."

Then Kirim Osmanov with the capsule flew outside the cave. He can't remember anything else about this event and is still wondering what has fully happened on that day. The truth about the aliens is one the most important things to mankind. People don't want to be alone in the universe. We are seeking contacts with similar beings and hoping they would be more advanced than us to solve all our problems and answer the question about the origin of life.

Furthermore the aliens have very good technology which means financial and military superiority. That's the reason why the governments don't want to expose their secrets about extraterrestrial civilizations.

Anton Anfalov: "Starting from 1941 until now there is a huge military-political conspiracy worldwide about concealing contacts with an extraterrestrial intelligence, concealing alien technology and concealing the contact with them and using the obtained knowledge in the interest of a small elite group that has actually imposed a monopoly on this."

But why are the aliens not interested in exposing information about themselves. Nobody has ever stopped them from landing on the front lawn of the White House. Or is their goal about our planet in fact the reason to hide their existence.

Information about UFO's is inaccessible even for the U.S. President and his administration. The elected representatives which change every four years are also unable to find out anything. In the state system of the United States there are dozens of levels of secrecy. The bulk of modern American leaders do not exceed more than eighteen levels and access to nuclear codes.

Anton Anfalov: "And in addition to that they shouldn't know anything new; you do not need to know this information even if you have all the access on this scale. You won't know anything even the existence of perfectly hidden 'black' and 'super black' program secrets on alien technology which, by the way, has existed for many years in the United States.

Information about alien technologies is possessed only by a small group of people know as a 'secret society' that rules the world. This society has agents not only in the United States. Vladimir Zamaroka almost all his life has worked in the aviation space sector and along with that he has observed anomalous phenomena. From his observations he has compiled an impressive file including information about the 'men in black'. He has even met one of them in person.

Since the time of 'Perestroika' he has travelled by train on distant research and telling all night stories about UFOs to his companion. One morning someone knocked on his door.

Vladimir Zamaroka: "It was man with a height of two meters maybe wearing a black hat, black suit and white shirt. He had a ticket to the adjoining compartment on the train but came into here for an unknown reason."

After participating in general conversation he suddenly asked if Vladimir and his companion would retell yesterday's story about the aliens.

Vladimir Zamaroka: "With an emphasis in speech and strange noise or something like that."

What has happened after that they both cannot remember. The man in black simply vanished. From that moment on and for a long time after, Vladimir Zamaroka discussed about flying saucers only to a small group of people. And a few years later as an editor for a science magazine he received a letter from a resident of Kaliningrad. In the letter a women described how she has seen many UFOs in her lifetime and she decided to go public with them. She sent him all her manuscripts in her letters and as soon as Vladimir got them, some interesting things happened.

Vladimir Zamaroka: "She was charged 13,60 kopeks for sending the two letters; sending these manuscripts had deprived her of having enough money to buy food. On her way back home after sending the letters, there suddenly appeared a young man in front of her. He was dressed in black. She was approached by this young man dressed in black and he said to her, 'could I give you some money?' His words were so polite so the woman took the money. And after he disappeared from view, she counted the money and realized there was 13,60 kopeks. Same as the expense for her letters. Why did the man in black her her?"

Today almost half of all population of the planet believes that the aliens exist and visit the earth. Twenty percent are convinced that aliens have bases in the USA. Such sustainable public opinion is possible thanks to the press and the movies that are constantly using this topic. It leaves one to wonder how much of this media coverage corresponds to reality? And why has so much information been put into the consciousness of the masses?

Alexander Semenov: "I must say that over the years, the number of these 'men in black' phenomenon has been reduced compared to the 1950's or 1960's including events and information. From the memories of the military personnel and declassified archives, it is known that U.S. special services actively manipulate the public opinion. The CIA had an initiative in 1953 where a whole set of measures were developed which include tactics about explaining all UFO cases as natural phenomenon. Later to mock the interest in flying saucers, radio and cinematography were extensively involved. Then when an uninformed man on the street became an eyewitness he was treated as a liar or just a crazy person."

The UFO witnesses are usually willing to remain silent to protect their careers or reputation. Sometimes tricks are used by the media to distort stories so that they are only half true.

Let take the example of Roswell in 1947. There was no UFO but there was a spy balloon called "Mobl". (The documentary at this point shows an example of film staging being done in a studio to recreate the look of aliens being examined.)

Martha Romonov: "I have met a commander of the Air Force in Alaska. He witnessed an alien crash. At the place of the accident he saw some corpses. Then a commission was created to investigate everyone involved."

Soviet ideologists showed a great interest to the people who were interested in UFOs following the examples in the U.S. The plot of 'silence' began in the

USSR starting from 1952, ridiculing and discrediting UFO issues. At the report dedicated to the 35th anniversary of the October Revolution, Mikhail Pervuhin as a member of the Presidium of the CPSU Central Committee had spoken about UFO reports. Then most of the American people lost their tranquility and started to observe the sky. Many of them have reported that they have seen strange things in the sky like huge flying saucers, flying spots and colored fireballs. But the UFO appearances were not only in the U.S.

Alexander Komechev: "With my flying experience I have seen a lot of flying objects especially in 1967."

In 1968 the newspaper 'Justice' totally discredited the UFO subjects. It had published a derogatory article about the flying saucers. The existence of the aliens was declared as lies and science fiction. After that this topic was considered closed.

Soviet 'men in black' and the agents who had this information knew their job quite well. Both sides rushed into a space race with a confidence that they will be there first and be the only single force in space. Among them in 1978 a secret UFO study program existed known as 'The Network'. During the 80's this program was called Galaxy and in 1985 it was renamed to Horizon. There they analyzed all the data that was collected earlier.

Further research about the flying saucers has led to shocking results. Military forces have discovered that their facilities have also become a subject to the UFOs. A quite familiar case about this is when above the 50th RVSN Division in the Carpathian District unexpectedly appeared several UFOs and activated the system of the nuclear missiles. It turned out that none of the operators could stop this command. All equipment was totally blocked. The control of the entire facility was regained at the moment when the UFOs disappeared. Provocation from the USA has not even been considered.

Starting from that moment the military seriously began to have fear for more powerful forces that originate from the depth of space. A secret study center was created in the city Mytisci near Moscow. It was a special laboratory at the 22nd Scientific Research Institute formed under the initiative of the USSR defense ministry.

Nickolai Antoshkin: "And it seems that much more advanced civilizations are somehow controlling us. We could conclude this by their appearance on areas where our nuclear or chemical bases are located; large hydro technical facilities and regions where the situation is unstable and in similar such places. Apparently they are controlling that we don't do anything bad on earth perhaps preventing what has happened on other planets in the past but I cannot imagine that at the moment."

Military aviation has many times received an order to destroy a UFO and obtain its fragments. But when the fighters reached near to it the attack on UFOs always ended badly. Colonel Alexander Kopeykin personally had to fly and intercept a UFO. In the summer of 1982 he had tried to get into an unknown black cloud that appeared over the military base. The diameter of the cloud was about one mile. It stood there totally motionless and was undetectable by radar. The contact with the airport over which this cloud formed was lost. The attempt of the pilot to get close to the cloud would almost cost him his life. The plane lost control and activated the alarm.

Kopeykin broke radio contact when the plane lost control and tried to get the plane away from the cloud. He managed to get away by some sort of miracle.

On another training flight Kopeykin's plane had approached another unknown object with clearly visible lights. The pilot came under a strong hypnosis.

Alexander Kopeykin: "Suddenly while flying toward the object literally I was paralyzed and when I moved away from there... very quickly I regained my consciousness."

Later Colonel Kopeykin was responsible for collecting such information in the southern sector of the Moscow Military District. The methodology of documenting UFO appearances belongs to him. The objects actually appeared as he was saying. The interval was exactly one month eleven days and two hours with a margin of error of ten to fifteen minutes.

The development in technologies has made a huge leap within the second half of this century. Such a rapid development of technology is very difficult to be explained simple from human's imagination of from market competition. It is quite certain that many products in the computer science or medical industry could not be made by the writers of science fiction or from the stories of UFO eyewitnesses.

Anton Anfalov: "In those inhabited facilities, the advanced civilizations transfer their own experience. They also transmit information to the most developed countries on Earth like USA, England, France and China."

Is it true that the most developed countries rely on the usage of advanced technology at the expense of its citizens? Is this just an eternal game? Or can humanity change these rules?

Studies made by ufologists and long term analysis have shown that humanity is frightened by the presence of UFOs. Flying saucers have been seen taking cubic meters of water and objects with lights illuminate the earth and some are extracting and removing minerals. In the USA there was a period when a lot of dead animals and slaughtered cows were found. Although on our planet there are millions of eyewitnesses and we have seen many activities in the night, still no one believes them. What are they getting in return? Wireless internet? Smart contact lenses? Nano robots and sensor displays? It seems everything they ask just so we could fly into space.

Anton Anfalov: "We will never know the truth. All I know is only crumbs. They are separated pieces of a large broker mirror. And I am collecting these pieces for many years but why? Because not only the military but also the politicians are not interested in letting people know the truth same as the extraterrestrial civilizations who are using them to drain our resources, materials and to get whatever they need."

Could all of this be created by non-human geniuses known only to a few people? Once over Tehran a UFO hung for several hours. Two Phantom Fighters took off into the air and started firing. They used all of their weapons but didn't cause any damage to the UFO.

If these beings are not aliens but a global myth then this got out of control a long time ago. The legends of grey people in flying saucers have been embedded into our consciousness a long time ago. They exist even in children's comics and classical literature. Not all the people believe but almost everyone has heard of them. Maybe that is how it's meant to be and 'men in black' and their motives are to remain in the shadows forever.

End of Transcript

The above transcript adds context to the comments made by the Russian President Dmitry Medvedev. The reader will have to determine on their own how much weight should be attributed to both his quote and the Russian documentary film. Together they paint an interesting picture of the possibility of aliens inhabiting earth as viewed from a Russian perspective.

It is one thing to hear hobbyists and casual observers developing theories about the existence of aliens but when comments are made by high ranking political or military leaders it is difficult not to sit up and take notice.

In my research I have tried to stay focused on information that is delivered by these high level sources or first person accounts. There is just so much information available on this topic that it is easy to get bogged in speculation from academics, anonymous sources, discussion forums, science fiction writers and even from one's own friends and family! Everyone seems to have their own unique opinions about the possibility of alien life and the existence of UFOs.

If any type of disclosure to the public is imminent; it would naturally have to be corroborated by and announced by government and military sources. Press conferences and news announcements have already been made by non-government sources and these, for the most part, have fallen on deaf ears or been ignored by the mainstream media and the general public.

Section 1 – Current Well Known Commentary

When compiling information to be included in this book I decided I needed to start with the existing popular evidence. Some items can be quickly dismissed while others take a little more time. I knew that I had received valuable evidence for the existence of Alien life so I was hoping that I would naturally find corroborating evidence in the public record. At first, I was quite disappointed since a lot of the testimonials I found very hard to ascribe any credibility to them. Here are some of the initial transcripts I found from public videos.

I should mention that I am not trying to "de-bunk" or "bash" anybody. I am only giving my honest opinion and since these individuals have stepped into the public spotlight, they no doubt are fully expecting to get a wide range of opinion on their views. So again, I do not want to come across as sounding malicious towards these individuals; on the contrary I actually admire them and commend them for their courage to be vocal and public on a topic that can so easily be subject to ridicule.

I should also mention briefly that I have not included any transcripts from Bob Lazar. This person permeates online documentation about UFOs and has some very entertaining comments; but since he has already been scrutinized by others[1], I will not mention him except in passing.

_____

[1] The reader can find details provided by UFO researcher Stanton Friedman at www.stantonfriedman.com under articles/The Bob Lazar Fraud

Comments from United States Army Sergeant - Clifford Stone

This transcript is taken from public video of a speech given May 2009 at the National Press Club in Washington D.C.

"Morning ladies and gentlemen my name is Clifford Stone I'm a Sergeant First Class with the United States Army. I had a secret clearance with nuclear ashority (sic); I could get the clearance that I needed to do whatever it was that was necessary for me to do at the time on special operations when I was called in on those...

What I am referring to here is that I was involved in situations where we actually did recoveries of crashed, of crashed saucers for lack of a better term and debris thereof. There were bodies that were involved with some these crashes and also some were alive. While we were doing all this we were telling the American public there was nothing to it, we were telling the world there was nothing to it. I'd like to go into detail on some of the cases about the nuts and bolts, bolts cases right here. ...

"But the whole situation is we set back we told the American people that there is no such thing as UFOs. I've been involved where we have recovered these objects we know them to be of extraterrestrials. In 1969 I had an event that happened to me when I while stationed at Fort Lee Virginia, we went to Indian Town Gap Pennsylvania. That would be my first exposure to any kind of we would be recovering an unidentified flying object. When we went there we already had people that were already in the facility we were a backup team which was supposed to be MVC because there was supposed to be some nuclear materials that was on board this craft."

"Later on most people involved were to be told that there was nothing onboard that it was nothing more than just a crash of one of our aircraft. I know better. Because I was one the people that approached it with a Geiger counter to get surface readings; I was the first person to go ahead and see that there were bodies on it. That would be the first of approximately twelve events. UFO crashes are not events that take place every day; they're rare. I know we're not alone in the universe. I know that the absence of evidence is not evidence of absence; it's evidence that's been denied to the American people..."

I stand before you today in my Almighty God and I tell you this, if Congress calls me in and says, "Will you testify in detail what you know?" I stand here today prepared and ready to do just that. Governments must never lie to the people for no reason. Thank you."

End of Video Transcript

When I first transcribed the above video clips of Clifford Stone I had misgiving about including it in this collection. He actually appeared to slur some of his words such as mentioning 'ashority' instead of authority and also stumbling over the phrase 'nuts and bolts'; however I decided to include his testimonial based on a couple of factors. The first is that I didn't find any compelling statements from people or persons attempting to debunk his statements and second because of his stated willingness to testify to anyone at any time.

Not only has his Testimony not been challenged but rather he has been welcomed into the group known as "The Disclosure Project" headed by Dr. Steven Greer. This sincere group has compiled massive amounts of documentation and attempts to pressure the government into releasing more information that they may be hiding concerning extra-terrestrials and UFOs.[2]

Later interviews of Clifford Stone seem to show him digressing into a complex reality; dealing with aliens in personal and minute details; discussing time travel and mentioning various psychic phenomena. He mentions "we had catalogued fifty-seven (57) different species..." and yet earlier he mentions only twelve (12) events. I can only speculate that his early experience in the Army may have triggered or caused a severe trauma to his psyche that caused him to 'go off the rails' a bit later on in life. However confused his later statements may have become; it is still worth noting his earlier testimonials as a matter of UFO disclosure history.

And this brings up an important consideration. People that have been in contact with alien life may have been exposed to a type of trauma which results in their perception and memory of past events becoming skewed. Sometimes the individuals display paranoid

---

[2] The reader can visit www.disclosureproject.org for details on their work and goals.

behavior, irrational fears and severe depression. One wonders whether there are effective methods of medical treatment such as hypnotic therapy or specific drug treatments to help these people who have been exposed to such life changing events.

In an interview dated April 17, 2009, Clifford Stone is shown working as a security guard; in fact the interview is actually conducted while he is doing his rounds in what looks like a school facility. He appears to be a humble and unpretentious individual. Here are some excerpts of comments he made in that interview: brackets () indicate there may be missing words or garbled words in the transcript:

"...when I was in the Armed Forces, I had many titles, cause you had old () job...personnel services information officer, unit non uh unit non commissioned officer for biological-chemical-radiological warfare; also we had uh ()common that I had to deal with. I was supposed to be a clerk-typist, although I went to school. I never took a day at school. However when I got out of school at Fort Jackson South Carolina, they said I could type sixty words a minute. I can't even do that today. I'm good if I get twelve.

It wasn't until, what was it, when in '68, it wouldn't be until '76, I'd rarely see a typewriter. | Well the real () poster shop that they had me work on was a computer. Then we had computer, a classified computer, uh that you got all your classified messages on, and that had to a trip in the hand () because you had to go ahead and verify the messages when they came in so you had a code book; you'd get say five or six, seven, eight, ten part message; then you'd have to decode it and authenticate.

But, what you really want to ask is 'what was my official title when I was involved with UFOs?' Uh, the truth of that is it wasn't given an official title. You were a nobody. In some of the cases I was involved with, you'd go ahead and take off all your insignia, all rank; you had the name tags and you had a U.S. Army tag on your fatigues; you even cut those off. They'd give you a little plastic bag; you just put everything in that plastic bag. For the most part, you didn't know anyone else, uh it was meant to be that way."

End of Transcript

And here is some excerpts from a more recent interview transcript of Clifford Stone conducted in 2012 with Ed Komarek (I have omitted interviewer comments for brevity)

"...as a child I had friends that no else could see; by virtue of having these friends that no one else could see, I thought everyone should be able to see them. But they would always caution me; 'only you can see us'. And they looked just like children; I mean, the same as you and I. It wasn't until one day, I looking ahead and I'm trying to help this little bird that fell out of its nest and it broke its beak. Well as a child, you go ahead, if you have a cut you hold it under water. Not knowing that killed the bird, I held the bird under water to try to stop the blood. So that bothered me, I cried over that little bird for better than a week. But then something strange happened; one of the children who later; how do I put this; one of the children- we became close, I got to see the physical appearance at that time and of course in a very startled way and I can tell you startled; I was scared I guess for lack of a better term.

Umm and this entity, no longer looking like a child was asking me, why do I feel the way I feel because it too feels exactly what I feel, 'why do I feel that way?' And of course I ran, I hid behind the refrigerator, I felt like a bony hand just scratching my head saying 'you can run but you can't hide' and finally, after the shock wore off, they were telling me not to be alarmed that they were just interested in our social structures and I told, I believe at the time, don't pay a book telling it, and I was told books don't tell you everything, the only way you really learn anything is through the life experiences of individuals.

With that being said, I haven't talk about this much, I think it's time cause this maybe one of the last interviews I give – our visitors will choose certain people, and they will follow those people through a lifetime. They'll question some of the feelings, thoughts, the emotions of that individual throughout that lifetime.

They won't interfere with that persons' life, but most people will very subtly just go through life and never even talk about it. You have people out there who have had experiences that to this day have not come forward and talked about it; otherwise they live very normal lives.

Like you know I've had UFO researchers-'well we could believe your story more if you would say that this entity that you have ongoing contact with, if you would say he was grey, three, three and a half foot tall,.." Well, that would be a lie.

The entity that I have contact with is approximately five foot two, five foot three, something along that range, has like a light green skin...about the closest thing I can compare it to; as a kid I was fascinated with lizards so it wasn't uncommon for me to always have a chameleon and the skin on a chameleon particularly it was green, that's the closest thing I can compare it to...uh, facial features were like dark eyes only, look like a tear drop, only the pint of the tear would be at the top, was pointed inward like in here (pointing at his face) had a small protrusion (pointing at his nose) not large, almost not noticeable and two little slits, the mouth, the mouth was just a slit, there was no noticeable lips; however they can express emotion, they can smile. I don't wanna just say holes, they had configuration there but no ears.... Oh absolutely, I found out later on, had the dinosaurs survived we might be dramatically different also....

I've experiences with supernatural events prior to that. When my aunt died, we had experiences with her coming back to the house....I'm bringing this up for a reason...you're gonna find that people who have these experiences, have psychic experiences and it scares them a lot of times because of religion. People feel that they have psychic occurrences – it's the devils work. Uh, people have psychic experiences on a daily basis; sometimes so minute that it's not noticeable; but I had experiences because prior to that I didn't believe in the supernatural neither. So I got to the point where I believed in the supernatural, so I got to accept that. Uh with this entity...physical form everything...it had three fingers and one thumb. They were long and slender. No noticeable nails or anything like that, but they were slender and an opposable thumb...

But the one thing I never did get uh, we go over what sets us apart from monkeys and for the longest of time, the entities name was Korona...spelt with a K I don't know why that's significant, K-O-R-O-N-A , I think that's how it's spelt, he said with a K, but he would always go like this (touching thumb to fingertips) and never answer and I said 'okay, if your people know about the evolution of earth, what sets us apart from the monkeys?' And I could tell he was smiling in his own way. He'd just do this (touching thumb to fingertips) better than a year that's the way he'd be doing it.

Then I was talking to a friend of mine that I met, which I'll get to him shortly, who was an Air Force Captain and I told him about this in our passing conversations; he says he just went like this huh (touching thumb to fingertips again) I says well remember he's only got three; and he says 'yeah but you know what this means?' No. I have no idea how this sets us apart...

I was about eight years old when I pointed this out. And he says it's simple, he's showing you something, monkeys can't do this, humans can. Opposable thumb. And it makes a big difference in our evolution...

Anyhow, Korona comes from a star system that's approximately a hundred light years from the planet Earth. It's within the Milky Way Galaxy however I'd caution anyone to be careful because the situation is, we're not just dealing with the universe, we're actually dealing with something they call multiverse. And in short within different space time, there's more than one universe and as our science progresses we will get to understand this and sometimes, not meaning to, but there'll be a distortion in the space time where we see things from the future from the past and even from other dimensional realities, sometimes we fall in there.

...to give you an example, and I always like to use it because I know it to be true, based on their technology,...they can go ahead, leave their star system, get to our planet, our star system within an hour and forty minutes real time, yet they've come a hundred light years.

...their lifespan is roughly three hundred years

...they actually, they have love in their culture...

...he had four children...

End of Transcript

The interview goes downhill from there, he discusses time travel, fractured souls, forbidden questions and so forth.

Comments from Retired Marine Core Captain and Test Pilot – Bill Uhouse

This transcript is from a public video of Retired Marine Bill Uhouse. He comes across as a very personable and likeable character that had interesting experiences in the military:

"We received the disc from extraterrestrials. This particular one that I had, had to work on, or did work on was from the Kingman crash and that was later on taken to a Nevada test site where at the time I didn't know. No bodies, there were four live feeling good guys. Four live feeling good extraterrestrial type individuals yes. Initially they were taken into Mexico and then several months later they were taken to the Nevada site.

One of the things I worked on was a flying disc simulator, trained pilots to fly this strange looking craft designed for humans to fly not for any extraterrestrials to fly.

J-Rod is a grey alien about 5'4"; his role was only as a translator, scientific translator that was all. If we wanted to put something into place that they had, he had to agree with it. I would have a question, you know, and I would bring it up in my mind just how I wanted to present it with him and he'd already know that I had this question and he already had the answer for me and if he responded it would be in my voice. And you wouldn't even open your lips. It is possible that he and a few thousand others are working on the project either here or elsewhere; a few thousand other greys like himself."

End of Transcript

In another interview, Uhouse starts the interview by stating, "I'm charging him for this..."[3] and then makes the comment "...there are only a few of them that are here..." so obviously it is contradictory for him to say it possible that there are a few thousand greys.

Later on, he goes into specific details about how the greys live. He states:

"...they only have one religion and it's not the same as we know it; theirs is a more or less spiritual type religion, they don't have a church they go to; they don't, you know, go to church every Sunday I mean nothing specific." At that point the interviewer asks him bluntly "How does one know this?" and Uhouse responds:

"How does one know that? Well you have to uh, go to a long period of, a of uh, I don't want to say

───────────────────────

[3] Dec 1st 1994 Interviewed by Miles Johnston

acclamation but it's a long period of a, of understanding. Understanding and knowing what questions they ask. And you know I'm trying to understand how I got it okay because that information; a lot of it is like uh you know asking me what they eat you know. Or what do they want from us; and that type of thing."

"I mean each one of those from those different planets could be different in what they really want from this planet. It could be like I don't understand the abduction scenario. I don't understand the various different so they speak the number of different aliens that are supposedly coming here. From what I understand there's maybe only no more than six different species if that and not the seventy or seventy three that John Lear talks about."

So again, you start with a situation that sounds reasonable and is even verifiable that an individual had a specific history usually in the military at a specific time and they encountered specific things but then at later times they go off into tangents about very obscure details about aliens which are impossible to verify and are sometimes even come across as nonsensical.

I will leave it to the reader to make their own opinions of these statements but for myself, it is very difficult to have confidence it stories that go off into very exact details about things that are basically impossible to verify. For example, I can accept the testimony of someone talking about their military work experience and I can take into consideration their opinion on what they may have seen at a specific time; but if they start talking about what daily life is like on Zeta-Reticulae as if they are talking about a trip to the local Walmart it causes a sharp disconnect in my ability to have any belief in what they are talking about.

There are other individuals that I won't mention here that talk about their experience with UFOs and then feel the need to go into a number of unrelated topics. These topics, however, they feel are very much related. For example you will find people who feel that UFOs are related to any number of the following:

1. Daily life on Zeta Reticulae
2. Daily life on Pleiades
3. Atlantis
4. 9/11
5. Kennedy
6. The Illuminati (or any secret society of your choosing)
7. One World Government

These people insist on intertwining these and a number of other topics into a hodge-podge of blurry visions that go in circles. As fascinating as they may be, they provide little help to anyone that is sincerely curious about UFOs and Aliens. I don't know why all these topics seem to morph into one another with respect to a number of curious people but that just seems to be the strange current state of affairs in this area of study.

Comments from Command Sergeant - Robert Dean

"...It began for me in 1963 was I was assigned to the Supreme Headquarters Allied Powers in Europe. I arrived in the summer of '63 and I was assigned to the Operations Division; I was further assigned to what we call SHOC the war room, the Supreme Headquarters Operations Center. And at that time I was given a cosmic top secret clearance which was and still is the highest level of security access that NATO has.

When I was assigned to the war room I learned that a study had been initiated in 1961 and everybody was talking about it, they were gossiping about it. And the subject was UFOs and that intrigued me a bit.

When that study was concluded in 1964 and it was published and I was still working in the war room at that time; I had access to this document, I had a chance to read it. And having been exposed to it I must tell you honestly that my life has never ever quite been the same. Because I read first hand reports, verifiable NATO military material that indicated that the UFOs were not only real but they represented something far beyond anything I had ever imagined before.

When the study was concluded in 64 they concluded that there were four different groups apparently coming and visiting us. Out of those four different groups, one group looked exactly like we do, so much so that they could sit beside you in a restaurant or in an airplane or in a theatre and you'd never know. And that particular point bothered the military guys a lot.

The point being that some of these people from somewhere could be walking up and down the corridors of say headquarters or they could be walking up and down the corridors of the Pentagon; one day at lunch a Lieutenant Colonel made the remark, he said "Geez do you realize they could even be in the White House?" And there was a little forced laughter at that point I remember because a lot of us over the years have had some misgivings about the occupants of the White House.

But the point I'm making is the idea that aliens could be here in our midst bothered the military because if you're not paranoid when you go into the service, you're certainly trained very quickly to be so; it's the nature of the military mind I think to be paranoid.

But that was only part of the story. You see conclusions in 64 there were four groups that we knew of. They were apparently interplanetary, some very likely were interstellar. They even concluded that some of them could be intergalactic. When I retired in 1976, many of our military people knew at that point that we were not simply with visitations from people from other planets or stars systems. They had concluded by 76 that some these visitors very well might be multidimensional in their source. The evidence that we had collected and the evidence that they had repeatedly demonstrated to us; and that's no accident. It became very clear after a time that there was a programs or a process of some kind under way that they had demonstrated over and over and over again that they apparently were able to manipulate matter and time. Now this really shook up our scientists.

There are a lot of new young scientists in particle physics today who talk about multiple dimensions. There's a young a brilliant your professor in New York by the name of Mitcheo Kaku who has written a brilliant book called Hyperspace where they talk about ten separate dimensions. I'm not enough of a physicist to try and explain to you what a dimension is but the idea that there could intelligences from somewhere else, from other dimensions coming and going into our reality. It's been quite a shake-up in traditional science.

I guess I speak out openly and bluntly about this because I feel so strongly about it. I violate my security oath every time I speak about it. I do it intentionally and I do it on purpose Because I feel so strongly that the American people not only have the right to know the truth but they have a need to know the truth.

The truth apparently is simply this that we're not alone, we've never been alone. We are apparently part of an infinite universe filled with intelligent life. I find that exciting. That doesn't frighten me. The shape study concluded in 64 that if they, whoever they were, were malevolent or hostile, they could have taken this planet and cleaned it up, eliminated us, turned us into dog food or whatever, a long, long time ago. That the historical evidence indicates that they've been with us a long time and I've concluded that we've had what I like to call an intimate inter-relationship with at least one of those groups; the group that looks exactly like we do.

And at that point I don't like the word alien. I don't think the term alien is appropriate at all here. I like to refer to them as family. We're related to them. I think they had a hand in our being here. And I think the time has come where we are about to meet our extended family. It's going to be very soon. And the people are not ready.

And one of the reasons I do speak out so bluntly and so openly is if in some small way I can help people get prepared for this. Because I believe that this reality, once we have accepted it, and understood it, and gone beyond the fear, will bring about an expansion of consciousness in the human race that will truly help us and prepare us to go out there and take our rightful place in that infinite community of life."

End of transcript

The above quotes by Robert Dean are quite powerful considering his career and the positions that he held in the military. It is clear from listening to his recorded testimonials and statements that he is very passionate about what he talks about with respect to aliens and UFO's.

In his later videos (aged 78 for example) he mentions that he was abducted and has actually been inside alien craft. His comments seem to follow those of Clifford Stone in that his comments made in later years are more expansive and detailed that those made in earlier years. Robert Dean even breaks down in tears in later videos due to his anguish in trying to spread his message; a message that in part details that humans are in a type of zoo as a type of genetic experiment carried out by aliens with technology a million years ahead of ours.

It is hard to see the motivation behind these types of grandiose beliefs. The financial considerations may contribute somewhat (as mentioned previously Clifford Stone is seen working as a security guard in one of his later videos) but I would suspect that both these men are relatively financially secure to the point where any financial gain would have little effect on their personal presentations. One might speculate that the mere possession of 'hidden knowledge' acts as a powerful mental opiate of sorts. The presenter gets a genuine psychological high in being the bearer of this almost mystical knowledge resulting in the ultimate level of condescension possible.

What may start out as single primary experience (Stone participating in the Roswell clean-up and Dean getting access to some high level government thesis papers on potential Alien activity) may cause the person experiencing such events to further experience a debilitating fracture in their psyche; the end result, after years of inner turmoil, brings on a definitional case of megalomania.

One can see the transition follow a remarkable parallel in both Dean and Stone who in earlier years speak as concerned ex-military citizens and gradually descend into a level of hyper-smugness with an acute and detailed knowledge of aspects of alien culture, quantum physics, all levels of government (even after many years of retirement), alien technology, religion, spirituality and the state of the consciousness of all humankind. This hyper-detailed level of knowledge causes them to be burdened souls; travelling the world with their message in the hopes of enlightening the few who will listen that humans need to prepare to accept this knowledge for their own betterment and the betterment of the world.

They weep crocodile tears over the sorry state of humanity that lies in ignorance of their own personal secret knowledge; they shed tears for the poor Alien civilizations that need our help to get back home or simply to understand them; the whole emotional soliloquy can be painfully difficult to watch.

In such cases as those similar to Dean one needs to separate the documented facts of the original experience from the resulting dissociative disorder that has developed over many years of frustration, paranoia and stress. The efforts that such individuals make toward spreading knowledge, however dysfunctional, should be commended and respected especially considering the amount of pain and suffering they may have been forced to endure at having been placed in a stressful public service situation in the first place that was originally not of their own choosing.

The downside of such instances of mental decay over time is that the latter testimonials of the individuals in question detract from the original factual events and experiences that may have actually occurred. These latter testimonials while intending to add validity to the original event end up having the opposite effect; merely becoming an exaggerated smokescreen of sorts.

Of course not all individuals experience such a similar pattern; many who have had alien and UFO involvement have maintained a consistent record of the events in question.

Comments from Military Intelligence Officer – Frank Kaufmann

This partial transcript is from a documentary interview published in 2000. I looked up the website listed for the production company and the domain is no longer active (has a link stating that the domain is for sale.) These comments deal with the Roswell incident and the reader may roll their eyes while thinking, "Oh no, not Roswell!" But I would suggest to keep an open mind as this case still provides one the most powerful firsthand accounts of alien contact. Here is the transcript:

"Well, let me first explain that this base in the 1940's Roswell Airfield was one the most important bases in the entire Air Force because of the many varied activities that were going on here. We had an ord bomb site here, it was one the training bases for the 509 bomb group that dropped the atomic bomb on Japan, and the fact too that you had experimental work going on at White Sands and also highly secretive work going on at Los Alamos in the northern part of the state so you could see this was a hot bed of activity; it was one the hottest places in the entire world.

Well in June of 1947 our radar people stationed at White Sands noticed a series of erratic movement of blips and on the night of July the 4th we were watching the blips; they must have been about 11:00 PM. There was severe thunderstorm and heavy rain going on throughout the entire area; the screen lit up. The radar man just couldn't quite make out just what was wrong; maybe there was something wrong with the radar screen but he was told it was in working order so he came to the conclusion that something went down east of White Sands; where (exactly) he did not know.

So in the meantime we were in constant communication, with White Sands, with Roswell, and the airfield with the base commander here and also the intelligence officer. We were ordered back to the base and on our way back we were still in radio communication. What caught our attention that something went down north of Roswell is that people driving on Highway, US Highway 280-85 north going south noticed an orange ball falling towards earth. Well it was common in those days to call the base and tell them that something unusual had happened say, so as we got back to the base; in the meantime Major Easley the Provost Marshall sent some MP's out on 285 to locate the, the, where this orange ball actually fell to earth and they noticed this orange glow off the highway about maybe fifteen miles inside; off the highway. And when we arrived back to the base, we were told to go out there and we had search lights and what have you, to the area, we had to click some wire fences, get into the area. And what we saw at that time was just unbelievable, it, it wasn't one of our craft we know what they look like, but what we saw was a craft of unknown origin.

***

Well after we got most of the craft and the bodies cleared out of the impact area and onto the base, the base hospital and to hangar 84 we convened in Colonel Blanchard's conference room uh to determine how we were going to handle the press you know. And there was quite few you know out there wanting some answers and, well how do you explain something like that you know; I mean it's something unheard of; so we didn't know anything about a UFO or a flying disc or anything that was extraterrestrial or what have you; see. So after, well after I think a couple of hours we finally came to the conclusion maybe, uh, well in fact it was Major Thomas who came up with the idea that maybe if we tell them the truth and take a chance whether they would believe it or not since this is something that is bizarre unheard of, see. So we agreed to that.

And it was rather interesting when we went to the, out of the conference room and met members of the media there, must have been about twenty-five or thirty and Thomas started off (laughs) very nicely and he said us, "Ladies and gentlemen", he said, "the announcement I have is that we have been invaded from outer space. We have a craft possibly extraterrestrial origin." When he said that when he said that word 'extraterrestrial' about two thirds of the group just bust out laughing and threw up their hands you know and said "the hell with this" you know and they knew we were gonna come up with some cockamamie story and they just left the room, so. But those that were there, there was a small group, they weren't satisfied, they wanted to know something about the extraterrestrials you know. Thomas says "well I don't know anything I'm just telling what it is you know, you asked me and I'm telling you; we haven't actually made a thorough examination of the craft or the bodies you know or anything of that nature", he said, "until such time we'll have to leave it as it is."

Well when we went back into the conference room, uh, Blanchard and even Major Easley the Provost

Marshall and Jessie Marcel the Intelligence Officer and the rest of us we said my God the plan worked, see, the truth you know, see, that's the way – we had no idea you know that the um, - it had nothing to do with what the paper came up with; the paper just came out with a flying disc you know was captured and of course the next day, Raimy came contradicting the flying disc story, you know and said it was a balloon, see. Well we knew what a balloon looks like, we knew what our planes looked like you know; but a craft of unknown origin is something else.

***

The craft itself I would say maybe might have been, I don't know, maybe twenty or twenty-five feet in length, maybe six feet in height possibly about fifteen feet maybe in width. There was a metallic smell you know around the craft. Inside the craft it had a kind of blueish-greenish light, see. There was quite a console out in the front and there was a little console in the back of the craft. I didn't go through the craft with a fine tooth comb, I didn't spend that much time, uh, checking out all the details, to say, because we had to get that, that damn thing outta there you know, before daylight. And when we got it on the flatbed, see, which we had one hell of a time getting it out of the area because of the mud you know sliding all over the place. We put a tarp over it and had part of the craft exposed. See, and we went right down main street with it. People looked at us you, there again it was a common sight; they would always see some craft going down main street, they looked at it, "huh another plane" something like that. See, we didn't want to hide anything. Then we went in the back entrance to the base, to hangar 84 and that where it was unloaded. Lickety split, no fanfare, no band (laughs) no flag waving or another of that nature.

That, that's you know you could say the art of deception. Act yourself, do everything as natural as what people are accustomed to seeing. But if you try to spend time hiding it they get curious.

***

Oh yeah they didn't have any of these slanted eyes or horny fingers you know or anything like that; they were, I don't know whether you wanna call them people. I call them people, that's what I'm accustomed to looking at is people see. They were good looking people, you know, fine skin, they were kind of ash-colored, the color of their skin. Eyes were just a little bit larger than ours, you know, more pronounced. Small nose, small mouth, very small ears; no hair; very fine features, well built maybe 5'4" – 5'5" in height. That's what I could remember seeing.

***

That in 1947 I can only put it this way the craft of unknown origin crashed north of Roswell; it was recovered with bodies not of this earth which proved that we are not alone in this vast universe of ours."

End of Transcript

Authors Note: The old VHS video entitled "Conspiracy-X Government Secrets Revealed" is no longer in production and I have not found a complete nor detailed transcript for this documentary anywhere. The quality and earnestness of the above testimony by Frank Kaufmann is compelling.

It is interesting to note that (unlike Stone, Uhouse, Dean, et al.) the credibility of Frank Kaufmann has been harshly attacked on a number of websites. Some even claim that he has already been exposed as a fraud. However upon close examination, it appears to me that his complete testimony has never been proven false. A document circulating online entitled "Frank Kaufmann Exposed" with the additional title: "Frank Kaufmann Reconsidered". The author of this research (Kevin Randle) (upon which the fraud accusations are mostly based) brings up a lot of questions but also makes the following clear statements:

"...he was certainly at Roswell in 1947 when the crash occurred, and thus could have been involved as he claimed."

"Frank seemed to speak from a position of personal knowledge. He didn't retreat into weasel words or hedge an answer. He was bold and confident in what he said and looked you right in the eye as he told his story."

"There were other good reasons to accept Kaufmann's stories as authentic. He provided details that seemed to dovetail nicely with other testimony."

"This timing suggested that Kaufmann's story had the ring of authenticity to it."

"These little things, which Kaufmann could not have known that we learned, suggested he was telling us the truth."

All the above quotes by Kevin Randle are from an article that is entitled Frank Kaufmann exposed? It would seem that the author of that article *wanted* to expose him but never actually found enough evidence to do so!

The co-author of the article, Marl Rodeghier found evidence that Frank Kaufmann may have tampered with or even forged some of his military service documents and so concludes:

> "Given all this evidence of counterfeit documents, we can have no confidence in any details of Kaufmann's testimony, even though he certainly was in Roswell in 1947 and worked at the base..."

> "The critical point is that we have determined the validity of Kaufmann's testimony, and can now discard it..."

Wow, that is an amazing feat of ignoring all the evidence in one swoop! That's what they mean when they say, "throwing out the baby with the bath water!" I wonder how many people (military or otherwise) have adjusted and/or modified their personal resumés at some time or another to make it appear more impressive? Does that give sufficient reason to discard everything else they ever may attest to?

The most compelling document in question associated with Frank Kaufmann is the Easley Memo for which Rodeghier and Randle claim to have found a copy. Here is a reproduced image of that very copy they mention:

Although very difficult to see clearly,(it was not possible to expand the image for this publishing format without significant degradation) it reads in part:

Headquarters

Roswell Army air Field

Subject: Recovery "Flying Discs"

To:

 As requested by () Memorandum dated () July () 1947.The following recommendation was made to issue a directive assigning a preliminary security classification and () for detailed study of this matter to () complete sets of all pertinent data which will be made available to AIND.

 For the purpose of analysis and evaluation, the craft recovered is assumed to be manned craft of unknown origin and may in fact represent an interplanetary craft of some kind. A detailed study of this matter to include the preparation of complete () of all available and pertinent data will be made available to AIRD.

---

[4] Some detractors have quoted the following document as the hoaxed Shulgen memo:

"For the purpose of analysis and evaluation of these so-called "flying saucers," the object sighted is being assumed to be a manned craft of unknown origin. While there remains the *possibility* of Russian manufacture, based on perspective thinking and actual accomplishments of the Germans, *it is the considered opinion of some elements that the object may in fact represent an interplanetary craft of some kind."* They then claim that the 'real' document reads as follows:

"For the purpose of analysis and evaluation of the so-called "flying saucer" phenomenon, the object sighted is being assumed to be a manned aircraft, of Russian origin, and based on the perspective thinking and actual accomplishments of the Germans."

This is another example of mis-information. They provide no copy of the memo; they only verbally mis-quote it. Similar to the "Frank Kaufmann Exposed" article, they quote Frank Kaufmann but there is no actual videotaped interview or transcript as has been reproduced above. The actual copy of the memo that is reproduced above makes no mention whatsoever of Germans or Russians.

The copy of Easley Memo above has the signature of Colonel Edwin Easley on it and neither Rodeghier nor Randle make any claims that it has been forged.

Another document that corroborates the testimony of Frank Kaufman is an FBI document that was released in 2011:

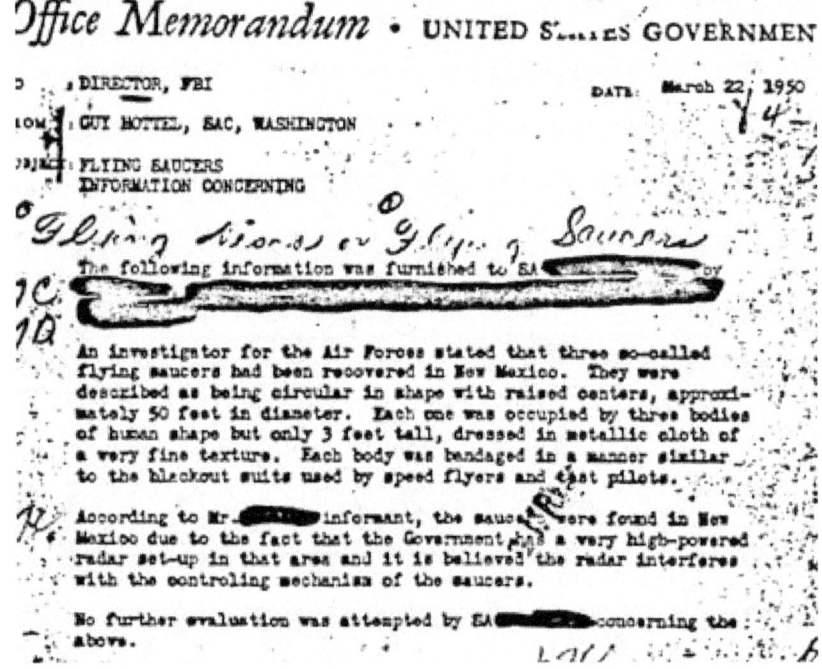

Office Memorandum • UNITED S....ES GOVERNMEN

TO : DIRECTOR, FBI   DATE: March 22, 1950

FROM : GUY HOTTEL, SAC, WASHINGTON

SUBJECT: FLYING SAUCERS
INFORMATION CONCERNING

An investigator for the Air Forces stated that three so-called flying saucers had been recovered in New Mexico. They were described as being circular in shape with raised centers, approximately 50 feet in diameter. Each one was occupied by three bodies of human shape but only 3 feet tall, dressed in metallic cloth of a very fine texture. Each body was bandaged in a manner similar to the blackout suits used by speed flyers and test pilots.

According to Mr. ____ informant, the saucers were found in New Mexico due to the fact that the Government has a very high-powered radar set-up in that area and it is believed the radar interferes with the controling mechanism of the saucers.

No further evaluation was attempted by SA ____ concerning the above.

At the time of writing is still available for download from the FBI website at:

http://vault.fbi.gov/hottel_guy/Guy Hottel Part 1 of 1/view

The memo is to the director of the FBI and is from Guy Hottel, the special agent in charge of the Washington field office in 1950. It states:

> *An investigator for the Air Forces stated that three so-called flying saucers had been recovered in New Mexico. They were described as being circular in shape with raised centers, approximately 50 feet in diameter. Each one was occupied by three bodies of human shape but only 3 feet tall, dressed in metallic cloth of a very fine texture. Each body was bandaged in manner similar to the blackout suits used by speed flyers and test pilots.*

> *According to Mr. () Informant the saucers were found in New Mexico due to the fact that the Government has a very high-powered radar set-up in that area and it is believed the radar interferes with the controlling mechanics of the saucers.*

No further evaluation was attempted by SA () concerning the above.

This memo is interesting in that it shows one arm of the government did not know what the other arm of the military was doing between 1947 and 1950. It is likely that the number of alien craft was mixed up with the number of actual aliens in the Guy Hottel memo.

There is still a ton of conflicting information available about Roswell in books, from videos, and of course from various websites online.[5] It is easy to get bogged down in details and arguments which I have tried to avoid here.

I do not agree with the efforts (by Randle and Rodeghier) to discredit the testimony of Frank Kaufmann and I believe that the Easley memo mentioned above is genuine. These two pieces of evidence are not easily dismissed and I will leave it to the reader to make their own judgment. Public videos of Frank Kaufmann are available online for the reader to evaluate as well.

Section 2 - Comments from Additional Sources

Comments from Australian Minister Reverend William Gill

William Gill gives details about an experience he had while working in Papua New Guinea.

"Can you imagine what it's like to look up in the sky and see a totally foreign looking object there just hovering, not very high up, maybe two or three hundred feet up in the air and glowing and two bipods jutting out from behind it, from underneath it and sparkling all around and some figures up there. This solid looking object and figures walking about on top and not the slightest noise whatsoever and so; we waved.

---

[5] At the time of writing, details of original Roswell news coverage is still available for download at http://www.angelfire.com/indie/anna_jones1/daily_record.html

Wouldn't it be wonderful if we could get this object down onto the playing field and as we waved wondering whether we would get some recognition and whether perhaps they would understand what we wanted, they waved back.

So I asked a boy to go quickly down, bring me a torch, bring me a pencil, bring me paper and uh return as quickly as you can so I can get any other events that occur and minute by minute movements so that at least we'd be able to talk about it the next day. And this he did, very, very quickly, he brought it back, but he brought the torch and put the torch on and shone it to the craft. And as he did so, he waved the, moved the torch this way (making a shallow u-shape with his hand) and were dumbfounded when we looked at the craft and the craft was as though it was responding to the torch, it began to do this too. (making a shallow u-shape back and forth with his hand) like a disc-shaped object just moving the same way responding to the movement of the torch.

Next day, just prior to the evening service about seven o'clock, the thing was there again, it had arrived about an hour earlier and we all decided to have the normal evening song that we do have on those nights, uh because well the thing was out there by the church anyway and we felt it wouldn't go away during the service and it didn't. When we came out there it was still up in the sky. And so for another hour or two we watched and then suddenly it did go. And there was this amazingly, incredible speed that the whole craft disappeared to nothing across the bay in the matter of a second or so.

Well now, what are we to think of this kind of phenomenon; people claiming to see things such as I did? There were thirty-eight of us and we all believe that we see it but of course we don't expect other people to believe us if they don't want to."

End of Transcript

Here are two illustrations that William Gill made from his experience:

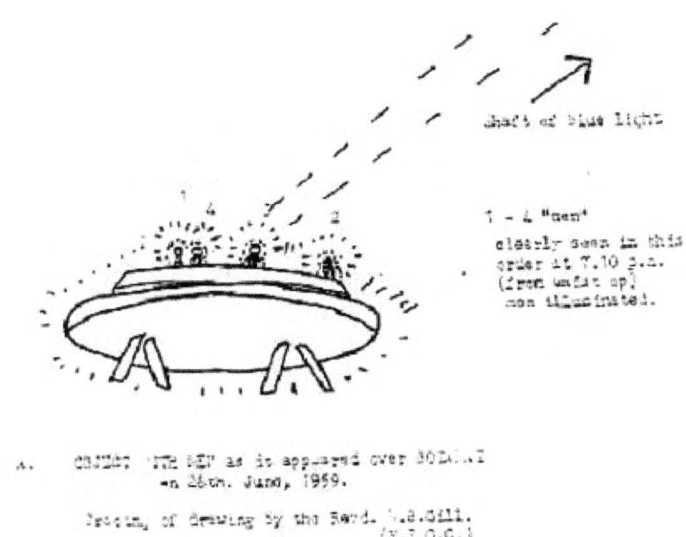

Shaft of blue light

1 - 4 "men"
clearly seen in this
order at 7.10 p.m.
(from waist up)
men illuminated.

A.   OBJECT 'THE MEN' as it appeared over BOIANAI
          on 26th. June, 1959.

     Tracing of drawing by the Revd. W.B.Gill.
                              (V.I.G.C.)

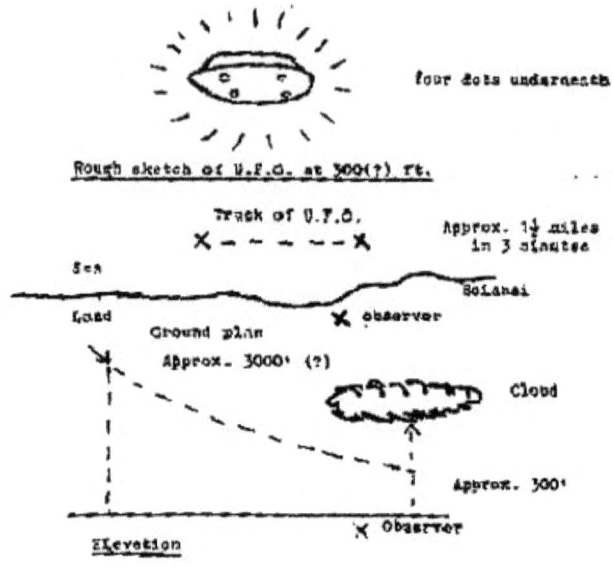

four dots underneath

Rough sketch of U.F.O. at 300(?) ft.

Track of U.F.O.
X – ~ – – ~ – X

Approx. 1½ miles
in 3 minutes

Sea

Boianai

Land

X observer

Ground plan
Approx. 3000' (?)

Cloud

Approx. 300'

X Observer

Elevation

His eyewitness account is quite remarkable for a number of reasons. One is that it was witnessed by many people; thirty-eight according to his account. Second, the sighting lasted for a long time; over an hour and on successive nights. And finally the point that the Aliens apparently responded to their observation is a very unique aspect of this report.

Unfortunately William Gill passed away in 2007 but there are videos of his lecture available online that the reader can view as a follow up.

Comments from NASA Astronaut - Edgar Mitchell

Edgar Mitchell was the 6th man to ever walk on the Moon. Here are some interesting quotes from him:

"Our destiny, in my opinion, and we might as well get started with it, is (to) become a part of the planetary community. ... We should be ready to reach out beyond our planet and beyond our solar system to find out what is really going on out there."

"...it's rather clear there's been a lot of misinformation, not just misinformation, there's been an active effort on the part of presumably people in the know to discourage the public participation, the media participation, to discredit the whole area. And that seems to have been true for at least fifty years. And also their existence of an extraterrestrial presence or an extraterrestrial existence even went against the traditional science of thirty or forty years ago; even though I went to the moon almost thirty years ago now; it was the conventional wisdom both in science and theology that we were alone in the universe and I don't think that people really believe that anymore; that was the conventional wisdom."

"...I happen to have grown up in... Roswell, New Mexico... where the Roswell incident of 1947 took place...but I have also been in military circles and intelligence circles that knows beneath the surface of what has been public knowledge that, yes we have been visited."

"…I don't know if it will be (a disclosure) this year in the United States but certainly we have already had it in the past few years, the Belgian government, the French government, the Brazilian government, the Mexican government, they've opened their files and admitted they had the files…"

"If you are interested enough to dig into it and want to know about it, you can know about it…"

This last quote really strikes at the crux of the matter; as well as his comments about misinformation. Because there hasn't been any effort on the part of the U.S. government to actively release truthful, helpful information, it leaves one with the task of having to wade through vast amounts of misinformation or even simply irrelevant information on the subject to find genuine facts of value towards knowing the truth.

Comments from Former Canadian Minister of Defense – Paul Hellyer

Paul Hellyer was a long serving Canadian Politician. He served as a member of Parliament (similar to the position of a Senator in the United States) from 1949 to 1974 and during that time was also in charge of the Canadian Military as the Minister of Defense from 1963 to 1968.

In April of 2008 he spoke at the X-Conference held in Washington D.C. Here is a transcript for a keynote address he gave at that conference. You can immediately sense the conviction and confidence that he has when addressing the audience. These are not the words of someone speculating about 'what might be' but rather he is speaking as an authority; one who is familiar with the subject and one who has been privy to information at very high levels.

"Decades ago visitors from other planets warned us about where we were headed and offered to help, but instead we, or at least some of us interpreted their visits as a threat and decided to shoot first and ask questions after. The inevitable result was that some of our planes were lost; but how many were due to retaliation and how many were due to our own stupidity is a moot point.

Wilbert Smith, one of the first Canadians to take an active interest in the subject of UFOs, asked the visitors about the accidental destruction of our aircraft by flying into the vicinity of a flying saucer. The response and this is Wilbert words:

"We were informed that although a few of our aircraft had come to an unfortunate end by what they considered the colossal stupidity of our pilots, they were now taking corrective measures to avoid our aircraft."

I asked them; and this is still Smith; I asked them what happened, and they said, "Well, the fields around the saucers in order to hold them up, in order to produce the gravity differential, the time field differentials which were necessary to operate the ship; these sometimes produced field combinations which reduced the strength of materials to the point where they were no longer strong enough to carry the loads that the materials were expected to carry".

Now, as we know, aircraft, particularly, the military type, are built with a rather small factor of safety, and in these regions of reduced binding, the materials are no longer strong enough to carry the load and the craft simply comes apart.

When I read that, I was reminded of the late great soprano Lily Pons who used to smash crystal glas-goblets with her high 'C'. She would direct the sound at them and something in the sound waves would change the structure of the glass and it would fly apart. Well this didn't satisfy our military chiefs who must have thought that it was more important to secure American nuclear superiority- even though using it would result in the annihilation of us all- than to take the hint and start moving the planet back from the brink of global disaster.

They, the military, must have been and still are, so paranoid that they feel it necessary to use the visitors' technology to fight them off, rather than welcome them as partners in development- though they may have seconded a few renegades to assist them in what can best be viewed as diabolical developments.

Steven has said that talking about UFO's is passé and that we should be talking or limiting our talk to exo-politics. Well Steven, I agree with you in theory, but in reality we have a problem when official U.S. policy insists that UFO's don't exist. The veil of secrecy must be lifted-and it has to be lifted now, before it is too late.

It is ironic that the U.S. would begin a devastating war-allegedly in search of weapons of mass destruction-when the most worrisome developments in this field are occurring in your own back yard. (applause)

It is ironic that the U.S. should be fighting monstrously expensive wars in Iraq and Afghanistan allegedly to bring democracy to those two countries when it, itself, can no longer legitimately claim to be called a democracy when trillions and I mean thousands of billions, of dollars have been spent on projects about which both the Congress and the Commander in Chief have been kept deliberately in the dark.

How much has been accomplished in sixty years of feverish activity by some of the most educated minds in the United States? Has America developed flying saucers that are visually indistinguishable from the visitors, as alleged? And if so, what do they propose to do with them?

Even more critical, what progress has been made in the development of clean energy sources that could conceivably replace fossil fuels and save the planet from becoming a veritable wasteland? Well who has the answers? Someone does, but apparently they aren't telling either secretaries of defense or presidents because they do not have quote "a need to know."

In a story told by Dr Stephen Greer, President Clinton was asked a question by White House reporter Sarah McClendon about why he didn't do something about disclosure. And Clinton replied, 'Sarah, there is a government inside the government and I don't control it'.

Excuse me. Doesn't the Commander in Chief and the person who allegedly has his finger on the nuclear trigger have a right to know what his subordinates are doing?

The people of the United States who have paid the bills have the right to know. The people of the world demand to know because it is our descendants, too, whose lives are in mortal danger.

It is time for the people of the United States to launch a new war against the evil of lies, deceit, and darkness, and go all out to win the victory of truth and transparency and light. Thank you."

End of Transcript

From the above speech, one may wonder who he was referring to when he spoke of those who were hostile towards aliens. He may have been referencing, at least partially, previous American authorities such as Ronald Reagan as can be seen from this quote: Ronald Reagan in September 21st, 1987 addressing the UN National Assembly:

"...perhaps we need some outside universal threat to make us recognize this common bound, I occasionally think how quickly our differences worldwide would vanish if we were facing an alien threat from outside this world..."

It is clear that Paul Hellyer does not see aliens as a threat but rather as an opportunity to learn and benefit from new technology and from cooperating with these alien beings. His statements are powerful and are coming from an individual with an impeccable record of public service. He can enjoy a peaceful retirement and has nothing to gain by making these public comments which are shockingly blunt and to the point.

Comments from Former President – Bill Clinton

Quote from a speech in Ireland in 1995:

"If the United States Air Force did recover Alien bodies, they didn't tell me about it either and I want to know."

Ten years later, on September 14th, 2005 Bill Clinton made these comments during a speech in Hong Kong:

"I don't know if you all remember this but there was actually when I was President, my second term, there was an Anniversary observance of Roswell, you remember that? People came to Roswell, New Mexico from all over the world. And uh, and there's also a site in Nevada where people are convinced that the government had buried a UFO and perhaps an alien deep underground because we wouldn't allow anybody to go there. And um... I can say now, 'cause it's now been released into the public domain....I actually had, I had so many people in my own administration that were convinced that Roswell was a fraud; that this place in Nevada was really serious, that there was an Alien artifact there.

So I actually sent somebody there to figure it out. And it was actually just a secret defense installation alas and just doing boring work that we didn't want anybody else to see. So I can't think of that. Let me; I'll give you a serious thing though. I spent a lot of – in 2000 I was able to participate with Tony Blair and representatives of the French, German, Japanese and Canadian governments in announcing that we had succeeded in sequencing the human genome and perhaps some of you have investments in all of these biotech companies and now you know how we cloned Dolly the sheep and now apparently they may have cloned a dog, we were all talking about this – my own view is that assuming we don't go and do something stupid like burn ourselves up with global warming or blow ourselves up with a military complex we could just as easily avoid.

I think a lot of these biotechnology issues will be the dominant sort of intellectual and ethical challenges of the lives of those of you who are ten, twenty, thirty years younger than I am. Uh, because I think that we are going to be able to save peoples' lives that, you know, my generation couldn't be saved and we are going to come up against the limits of our own mortality in a way that we never could before.

And a lot of the things that happen good and bad will be stranger than anything even written in science fiction. But I don't know the answers; which is one reason I'd like to live to be a hundred just to see what happens...().there's absolutely no risk of that given my misspent youth; I'm lucky to be here now!

If there is one (a list of secrets) then I don't know it. Really the Roswell thing I think really was an illusion; I don't think it happened; I mean I think there are rational explanations that I have succeeded and I did attempt to find out if there were any secret government documents that revealed things; and if there were, then they were concealed from me too.

And I wouldn't be the first president that underlings have lied to, or that career bureaucrats have waited out. But there may be some career person sitting around somewhere, hiding these dark secrets, even from elected presidents. But if so, they successfully eluded me...and I'm almost embarrassed to tell you I did try to find out.

I do believe by the way, I'll be just as – one more flaky thing – you can also be flaky when you're out of office – I believe that now that we know there are not hundreds not millions but billions of other solar systems out there; thanks to the Hubble telescope ad what we know about Black Holes and the universe and all of that. The dimensions of Physics are such that I would be quite surprised if in the lifetime of people that are no older than thirty here, that we don't discover some form of life in another universe. It's pretty clear that there was something approaching elemental life on Mars at one time in the past based on what we've already discovered there.

So I say that only to say this, I hope all of you wherever you live will continue to support space exploration. Whether manned or unmanned is not so important but we have to keep doing that and I'm afraid that there will be a waning interest in it in the future. I think it's a great mistake; I think we should continue to explore the boundaries of our existence both end of the Earth and beyond the skies.

When I was President we discovered in the bottom of the Amazon River - we were just a small part of this but — two previously undiscovered forms of marine life so deep in the Amazon that they had never been found; and all the efforts of marine biologists.

So I think there are lots of interesting discoveries; biological, on Earth and other discoveries in the heavens that those of you who are younger will get to see unfold. You'll have all kind of problems with them but on balance it'll be a plus and it'll make life much more interesting."

It is interesting to note how Clinton's comments changed from his "wanting to know" to saying he thinks it was "an illusion". He must have been exposed to some inside information to dissuade him from pursuing the facts. In fact he was clearly more interested in biological developments at that time. He does however make the most powerful startling and powerful statement: "I wouldn't be the first president that underlings have lied to, or that career bureaucrats have waited out... there may be some career person sitting around somewhere, hiding these dark secrets..."

Perhaps it can best be summed up by the quote Bill Clinton made to senior White House reporter Sarah McClendon in reply to why he wasn't doing anything about UFO disclosure... "There's a government inside the government and I don't control it."[6]

There is also an interesting quote from Bill Clinton's Chief of Staff John Podesta on November 14th, 2007: "I think it's time to open the books, uh on questions that have remained in the dark on the question of Government investigation of UFOs. It's time to find out what the truth really is out there. We ought to do it really because it is right, we ought to do it because the American people quite frankly can handle the truth and we ought to do it because it's the law." So this sentiment on "wanting to know" is still present within the office of the President of the United States.

So one would naturally ask, who could be this "career person" or "career people" that would be hiding "dark secrets" if there was such a person?

---

[6] This quote by Sarah McClendon is attributed to Dr. Steven Greer who is head of The Disclosure Project and since Sarah McClendon has passed away, it is impossible to verify this quote from video transcripts.

The most obvious answer would be someone in the military. Let's look at some very basic connect-the-dots thinking.

General Curtis LeMay was in charge of Strategic Air Command from 1948 until 1957. Here is the transcript from some comments made by former Senator Barry Goldwater about a conversation he had with General Curtis Lemay:

Comments from Former Senator – Barry Goldwater

"I think the government does know; I can't back that up but I think that uh at Wright Patterson Field, if you could get into certain places, you'd find out what the Air Force and the Government knows about UFOs. Reportedly a spaceship landed. It was all hushed up, quieted and nobody ever, have never heard about much of it. I called Curtis Lemay and I said "General uh I know we have a room at Wright Patterson where you put all this secret stuff; can I go in there?" I've never heard him get mad but he got madder than hell at me, cussed me out, said don't ever ask me that question again."

End of Transcript

The above comments from Barry Goldwater were transcribed from video so there can be no mistaking his comments here. It is obvious that he felt Lemay was hiding something, and of course was sensitive to the question.

Lemay's career spanned from 1928 to 1965. He was in the highest levels of the military from WWII to the Cuban Missile Crisis with President Kennedy. In 1947 he was transferred to Europe to be in charge of the Air Force there and was stationed at Wiesbaden, Germany but of course we know that the Roswell incident occurred in July of 1947 so at that time Lemay would likely have been stationed in Washington D.C. at the Pentagon.

Section 3 - Military Control of Secrets

If the highest political leaders, i.e. the President of the United States, cannot provide, let alone discover for themselves, any facts about the validity of Roswell for example, then it would follow that people should be asking questions from military leaders and not political leaders.

Curtis Lemay passed away in 1990 and Strategic Air Command was absorbed into other military division after 1992. One of the military divisions that absorbed parts of Strategic Air Command is called Air Force Space Command.

Let's take a closer look at U.S. Military division called Air Force Space Command. Many people have never heard of such a division before and they have really only existed since 1982.

First of all, their insignia makes me laugh! I guess somebody in power was a real Star Trek fan! Here is the Star Trek logo:

Notice any similarity? Okay, let's leave the outright humor and silliness aside for a second. Air Force Space Command (AFSC) is basically in charge of any war activities related to space or cyberspace.

Some of the things AFSC controls and is in charge of include:

GPS (Global Positioning Satellite Technology)

Solid State Phase Array (Radar Detection System) at Thule AFB in Greenland

MILSTAR Satellite

The current head of Air Force Space Command is General William H. Shelton.

Here is a brief look at his military history from public records:

1. August 1976 - May 1979, Launch Facilities Manager, Launch Director and Technical Assistant to the Commander, Space and Missile Test Center, Vandenberg

AFB, Calif.

2. May 1979 - December 1980, Graduate Student, U.S. Air Force Institute of Technology, Wright-Patterson AFB, Ohio

3. January 1981 - July 1985, Space Shuttle Flight Controller, Johnson Space Center, Houston, Texas

4. July 1985 - January 1986, Student, Armed Forces Staff College, Norfolk, Va.

5. January 1986 - July 1988, Staff officer, Deputy Chief of Staff for Operations, Air Force Space Command, Peterson AFB, Colo.

6. August 1988 - August 1990, Staff Officer, Office of Space Plans and Policy, Office of the Secretary of the Air Force, Washington, D.C.

7. August 1990 - June 1992, Commander, 2nd Space Operations Squadron, Falcon AFB, Colo.

8. June 1992 - June 1993, Executive Officer to the Vice Commander, Air Force Space Command, Peterson AFB, Colo.

9. June 1993 - July 1994, Commander, 50th Operations Group, Falcon AFB, Colo.

10. August 1994 - June 1995, Student, National War College, Fort Lesley J. McNair, Washington, D.C.

11. June 1995 - September 1997, Deputy Program Manager and Executive Assistant, Cooperative Threat Reduction Program Office, Office of the Assistant to the Secretary of Defense for Nuclear, Chemical and Biological Defense Programs, Washington, D.C.

12. September 1997 - August 1999, Commander, 90th Space Wing, Francis E. Warren AFB, Wyo.

13. September 1999 - July 2000, Chief, Space Superiority Division, Office of the Deputy Chief of Staff for Plans and Programs, Headquarters U.S. Air Force, Washington, D.C.

14. July 2000 - November 2000, Director of Manpower and Organization, Office of the Deputy Chief of Staff for Plans and Programs, Headquarters U.S. Air Force, Washington, D.C.

15. November 2000 - May 2002, Director of Requirements, Headquarters Air Force Space Command, Peterson AFB, Colo.

16. June 2002 - January 2003, Director of Plans and Programs, Headquarters AFSPC, Peterson AFB, Colo.

17. January 2003 - May 2003, Director, Air and Space Operations, Headquarters AFSPC, Peterson AFB, Colo.

18. June 2003 - January 2005, Director of Capability and Resource Integration (J8), USSTRATCOM, Offutt AFB, Neb.

19. January 2005 - May 2005, Director of Plans and Policy (J5), USSTRATCOM, Offutt AFB, Neb.

20. May 2005 - December 2008, Commander, 14th Air Force (Air Forces Strategic), AFSPC, and Commander, Joint Functional Component Command for Space, USSTRATCOM, Vandenberg AFB, Calif.

21. December 2008 - July 2009, Chief of Warfighting Integration and Chief Information Officer, Office of the Secretary of the Air Force, the Pentagon, Washington, D.C.

22. July 2009 - January 2011, Assistant Vice Chief of Staff and Director, Air Staff, U.S. Air Force, the Pentagon, Washington, D.C.

23. January 2011 - present, Commander, Air Force Space Command, Peterson AFB, Colo.

That's over 25 years of distinguished service in a wide range of capacity and duties. That surpasses any politician with respect to access to knowledge about Space or anything that might come from Space.

Instead of listening to testimony from retired army file clerks and aircraft mechanics; people should be directing requests of information from someone like General William Shelton. A person in that capacity is in the highest echelon of military access to information.

There would need to be cooperation between the government leaders and the military to provide information to a publicly accessible database for investigation. Apparently this is the strategy that is being followed by some other countries that have opened Government resources to UFO documentation.

It is unlikely that Air Force Space Command would disclose any information themselves for a number of reasons. First, their obligation is to the security of the United States, not to helping the public find out about extra-terrestrials. Second they are likely too busy analyzing earth-based threats (such as nuclear detonations) to have the incentive to chase down UFO sightings.

Someone like General Shelton could assign a task force of some type to assist in helping the public have access to UFO information when sightings occur. With the resources available from Air Force Space Command, such a task group would be able to track and categorize a large number of UFO sightings. (The majority of which could be explained by earth-bound causes. There really hasn't been a formal group within the Government to address this need since Project Blue Book was dissolved in 1969). (There have been symposiums and some discussions but little that has had access to cutting edge technology like satellites and tracking radar.)

It is only with the help of the Military that progress can be made towards full disclosure of UFO phenomenon. These are the type of people that should be contacted and strongly urged to cooperate with civilian efforts to understand specific UFO reports like Stephenville or Phoenix Lights.

With such an autonous and unco-operative military; what's next? This:

Sure it's a joke but it makes you wonder who is behind the powerful military power-groups that seem to be accountable to no one but themselves.

Section 4 – Compelling New Evidence Released in 2013

Details on how these photographs were obtained

As I mentioned in the Preface, I met an individual in the late 1980's. If my memory serves me correctly, it was in the spring of 1989. I was doing what one might consider being 'mild' research on the topic of UFOs and had met a retired military man through a mutual friend. He, in turn, introduced me to this fellow who I will call 'Jeff'. I gave Jeff a call and left him a voice-mail; basically I asked if we could meet to talk about my passing interest in Aliens and UFOs.

I did not get a reply but after checking with my contact from the military, he assured me to keep trying as that was his normal response. I ended up leaving three or four voice mails on his answering machine, each time I would tell a bit more about myself and how I was hoping I could buy him lunch or if he was willing to meet for a coffee. I would have not have called a second time except my contact had said that it would be worth my while to meet him. So I ended up leaving a number of messages and then simply gave up and forgot about it.

Eventually, probably about two weeks after my first call I finally did get a call back. When he called, I remember I almost didn't know who it was and then it finally clicked in; I might have accidentally hung up thinking it was a wrong number. Anyway, Jeff did not acknowledge anything unusual about not returning my calls simply that he had been preoccupied with other things (maybe he was travelling, I have no idea, but it did seem strange that he took so long to call me back) and he agreed to meet me for lunch.

The lunch appointment was again strange it that Jeff was very late, probably about twenty minutes late, it was at the point where I was about to leave when he showed up and introduced himself. I had told him what I looked and he did not tell me anything about what he would be wearing so I had to wait until he approached me. Maybe he even waited until it looked like I was ready to leave for all I know but as soon as I did meet him, it did not take long to determine that he was suffering from paranoia.

From hearing, seeing and listening to a number of people on the subject of Aliens, I can't help but notice the incidence of paranoia to be much higher than one would find in the general population, and I guess that is to be expected. If a person feels they have had contact with an other-worldly entity that they do not understand, I guess it is only natural that they would be subject to a higher than average level of stress, fear and uncertainty.

Jeff was constantly fidgeting and looking over his shoulder to the exit of the restaurant or out the window, it was next to impossible to maintain eye contact or to build up any kind of rapport. I spend most of the time talking about myself, my background, my UFO experience as a teenager, my ongoing studies in the paranormal, my life in general, I even chatted about my then-current girlfriend if I recall.

Finally after about a half hour of one-sided conversation with me babbling and Jeff fidgeting, he finally let loose and opened up. He started speaking with a fast yet quiet voice, a steady stream of mumbo-jumbo; being under surveillance, having his food poisoned, suffering nightmares, getting the shakes and cold chills. He looked okay but from the sound of how he was describing things, he was really in bad shape. I was a bit taken aback and could only suggest that he see a doctor. He told me that he couldn't go to a hospital until he could get a new identity or "they" would find him and "take him away or worse". Jeff pulled out a manila envelope that he had curved up into his overcoat pocket and passed it to me. He finally made serious eye-contact and demanded that I only release the information if something "happened to him". Apparently he wanted to make sure that the information would be released to the public but was trying to keep it secret so that unseen forces would leave him alone.

Looking back I regret not taking notes or having a tape recorder with me. He had all types of minute details about how he was being monitored all the time, in all honesty I must tell you that a lot of it sounded very exaggerated; perhaps even like he may have been on medication that was causing him some delusional side-effects. He wanted me to keep the information secret until December 21st, 2012 which he said would be the 'end of all things'.

Over the next few months I spoke to Jeff a few times to ask him about the contents of the envelope but I eventually lost contact with him. For many months I did not look at the manila envelope. I eventually left it in a locked briefcase which in turn was placed in a storage trunk. Over time it ended up in the attic of my ex-wife's home for many years. (This storage of course was after I had moved on, divorced and remarried) I wondered about it often and knew it contained photographs.

Several times over the years I would pester my ex about getting the trunk back and there was always an excuse why I couldn't get it or when I did get access to the attic, I could never find it.

Late last year, my ex was finally moving out of the old home and asked me if there was anything I wanted to get from the garage and attic that was still mine. I went for a last look to collect some old books tools and to look for the trunk once more. I actually found it in the loft of the garage, not the attic. Someone at some time must have moved it out of the attic and stored in the garage since I never remember moving it. I don't know if anyone had been in it but the locked briefcase was still inside.

I had long forgotten the combination lock but it was a simple three digit number so it did not take me long to open it. (Seems the combination was 7-7-7; guess I was not very creative!) Sure enough, there was the manila envelope with the documentation, photographs, personal notes drawings and scribbles.

New Alien and UFO Photographs

Here are copies of the photographs that were in the envelope:

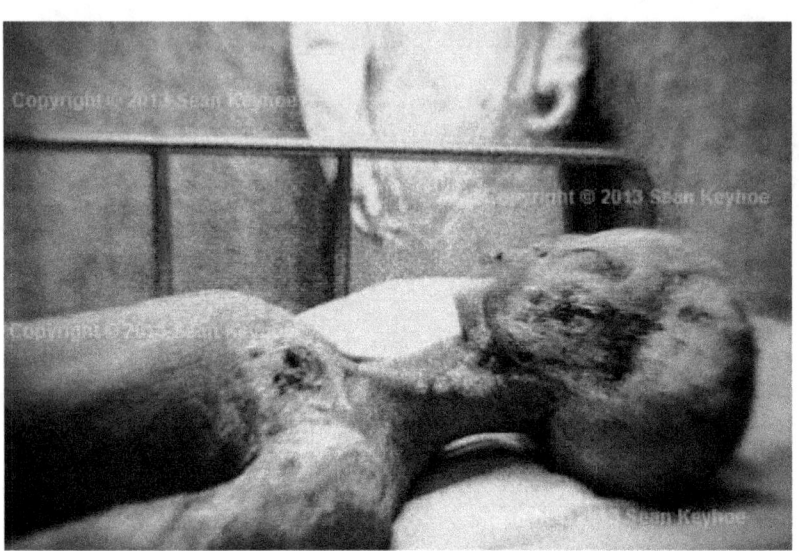

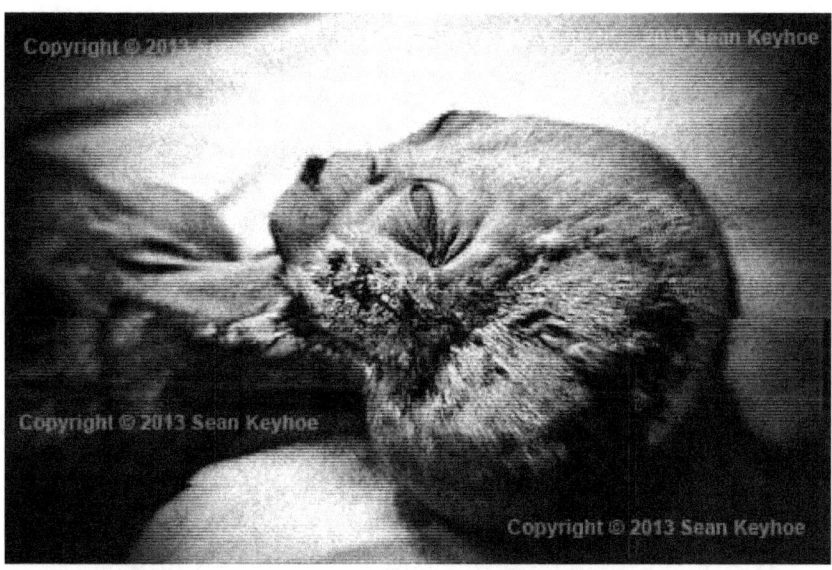

Jeff had told me that these were photographs of Aliens that were examined at Wright Patterson Air Force Base. They had been retrieved from a crash site and moved there for analysis. I asked him if there were of the same Alien or if they were two different Aliens. He said he did not know. I asked him if they were Aliens from Roswell. Again he said he did not know.

They appear to me to have either suffered burns or perhaps had been frozen, causing decay and/or damage to the skin. Beyond that I have no idea where these Aliens came from nor under what circumstances were the photographs taken.

In publishing these photographs at this time, I am hoping that a number of things may occur.

1. Perhaps a former medical worker from that era will come forward and recognize themselves in the first photograph.

2. A photographic expert will be able to determine if these are genuine or if "Jeff" was duped by being given fake images from somewhere.

3. Someone will come forward with additional images that match these and thus confirm them as genuine images of Aliens.

4. Someone will have the negative of these images and confirm the source.

5. Someone will recognize them as being photographs that they actually took; thus they could give the exact date, time and circumstances surrounding these images.

6. Someone will recognize these as being recreations from a movie production that was created in the past.

7. Someone will recognize these from a different location other than Wright Patterson AFB and

be able to confirm the original source such as location and date of the UFO crash.

8. A representative from the U.S. Military will request the return of the original photographs so that I will have a point of contact with the proper authorities.

If there is someone out there who claims to be the actual photographer, they can contact me and identify an item that has been cropped from one of the original photographs as a proof. These are scanned images of the originals so the resolution has been optimized for the Amazon publishing platform and they have been watermarked with copyright. Requests can be made for higher resolution images of the originals without the copyright watermark from the author if that would be helpful for ufology research.

seankeyhoe@gmail.com

Here are the additional photographs that were included in the envelope:

The above photograph had the following note folded over it:

1948 – This will have to be given to the correct people.

He is from Project Sign and will take care of all the details.

This next photograph did not have any accompanying notes:

If any readers out there can help me identify where these photographs were taken I would be most appreciative. The second image above clearly appears to be somewhere in Nevada, Arizona or New Mexico but I have never been to those places so perhaps someone can correct me. The distinctive rock formation should make it easy for a reader out there to pinpoint the location.

The previous photo above (the one with the flat landscape and the UFO in the upper right corner) looks much more non-descript and it even looks like that may be snow in the photograph, perhaps it was taken in a northern region.

Again, by releasing these images, I am hopeful that someone can verify the photographs or perhaps there is an expert that can use handwriting analysis to determine the origin of the note.

There were a few other inconsequential notes included in the envelope. These were the items that jumped out at me and I felt were of utmost importance to share with the public as quickly as possible.

I know that this information will likely upset many people; that is often the case with any new information on this topic. I have no interest in public infamy nor do I expect serious financial gain besides what modest proceeds may come from this book I would simply use to continue research in the area of Ufology. As a quiet retired citizen I have no interest or inclination to stir up new conspiracy theories, accuse the Government of new nastiness, nor to start crying for these poor Aliens because they did not receive a proper burial, as unfortunate or sad as that may be. I have had no telepathic experiences nor have I ever been abducted by anyone and I don't intend to start having such adventures now or ever.

If the government will contact me directly I will gladly return the originals when presented with a formal request.

Section 5 – Criteria for Proof

As much as I am excited to present this new evidence to the public, I must confess my own personal opinion on most new evidence that is appearing on a daily basis.

I feel that new visual evidence on UFOs will have little or no value for two main reasons.

1. The use of new flying technology will be indistinguishable from crafts that may have Alien origin. Organizations like NASA, U.S. Customs and Border Protection have started using aircraft-sized drones for monitoring ground based activity. Here is an example of just such a drone:

These drones will be able to hover and will likely have all sorts of flashing lights. Some will appear to be silent and some will be able to maneuver at crazy angles. In essence they will be indistinguishable from what people commonly call UFOs.

2. It appears there is an increase in the ability for individuals to produce faked movies of UFOs. That is, anyone with the necessary software tools can produce a video of a "Flying Saucer" on a desktop computer. Take a look at Youtube videos for examples of how fake videos are

becoming more and more sophisticated. If someone can make a video of a baby being scooped up by an eagle (one I saw recently) what is to stop them from making videos of UFOs abducting people?

There must be a move away from eyewitness reports or even video reports as the criteria for proof. The Disclosure Project headed by Steven M. Greer, although commendable in its efforts, has leaned too heavily on eyewitness reports; indeed there has been a steady stream of one witness after another which does contribute to tip the scales of evidence but all of it would be dramatically impacted by a single physical Alien space craft or a single body of an extraterrestrial.

Even if a complete Alien body no longer exists, it could be that someone, somewhere; perhaps a medical worker from the military that worked on the Aliens; knows of some remaining physical samples of tissues from the Aliens. This would of course be invaluable towards conclusive evidence acceptable to the scientific community and to the public at large. Hopefully if such evidence exists, it will be brought forward sooner rather than later.

Compelling Encounters with UFOs

On January 8th, 2008 a large number of individuals witnessed a UFO or rather UFOs over Stephenville, Texas. This incident was remarkable for a number of reasons.

1. The number of independent witnesses to the event.
2. The integrity of the witnesses. (Law enforcement officers etc.)
3. Supporting radar data to the event.
4. The object(s) were silent
5. The speed & acceleration & deceleration of the object(s)
6. Military involvement. Chased by fighters. (Subsequently denied.)

I will not go into all the details of this specific case but I do recommend the reader to examine the complete MUFON report and the MUFON conclusions that I have re-produced in the appendix. I am including a copy of the conclusions in the Appendix in case the original gets removed by the government for reasons of 'national security'. If enough people will read the documented facts of this case, the pressure may increase to the point where action will have to be taken to force the military to respond to and finally address the truth behind these unexplained events.

Summary and Conclusion

Since the 1940's(and perhaps prior) many UFO and alien encounters have been recorded verbally and then transcribed into written documents. Moving forward one would expect to see an increase in video recorded alien encounters. The explosion of 'video everywhere' brings its own challenges as videos are being faked.

The most compelling revelations will come with a combination of eyewitness accounts backed up by multiple video recordings. This would turn a page in the history of the world. Just as 1969 marked a milestone with humankind's first trip to the moon, the first live news broadcast of an alien visitation will likely mark a significant milestone in the history of the world as well.

The movie 'Close Encounters of the Third Kind' attempted to paint a picture of the first significant contact with extraterrestrials and scores of other movies have presented their own version of the possibility of such an event. Some producers present an image of benevolence such as 'The Abyss' or 'K-Pax'. Others portray a horrific situation such as 'Alien' or 'Independence Day'. And still many others like to simply portray a silly comedic spin on such things.

Whether one has seen any of these movies or not is not important; suffice to say that in both arts and literature; people have tried to imagine what alien contact will be like; they imagine a situation and build a story around their vision. The concept of alien contact itself has ingrained itself into the popular culture through all types of media. Books, movies, television and radio; all present possibilities and variations of different types of alien contact on a regular basis. One may expect that a collective sigh of relief will be felt when there finally occurs a definitive interaction with alien beings that can be understood and adapted to the benefit of all.

Some feel that such an encounter will pose a threat to humankind; such as Stephen Hawking who has warned that such an event could pose great risks to the inhabitants of earth. Others feel it will be an opportunity for mutual gain. It is no surprise that the extremes of both views are presented in the popular media as mentioned above.

Think of us as humans exploring other planets. How often have we visited Mars? A few times. How often have we crashed on Mars? - did we ever crash on Mars? Yes, we did - twice actually! The Mars 2 Mission from the Soviet Union crashed into the surface of Mars in 1971 and the Mars Climate Orbiter from the United States (NASA) crashed into the surface of Mars in 1999. That is two crashes out of over forty (40) missions to Mars with various Space Craft over a fifty year period (since 1960). (Most missions were not trying to land on Mars; but rather were orbiting missions or fly-by missions)

So you may want to ask the question, if Aliens have sent Space Craft to visit Earth, how many of them would we expect to crash? Would the ratio be similar? Less?(lower ratio due to their superior technology?) Out of the countless reported sightings throughout history, what if only forty or fifty were genuine visitations from extraterrestrials and out of those, what if only two or three actually ever crashed into the Earth? One of those crashes was likely what happened at Roswell. This kind of ratio makes logical sense to me. Can I prove this? No of course not; it is simply a logical mental exercise to compare our exploration efforts and results to the efforts and results of other possible civilizations.

One of the areas that was exciting to Bill Clinton, was the discovery of new species of life in the Amazon. How different would it be to discover a new species of chimpanzee in the Amazon jungle. Chimpanzees are humanoid, they are very clever. How different would it be to discover humanoid creatures on other planets that are clever and like to explore?

Only time will reveal how these events will unfold. The definitive proof of Alien life must come from some form of DNA confirmation. Many of the people who have had first-hand experience with UFO or Aliens have credibility issues and those that are credible are missing vital hard physical evidence to back up their claims.

If there existed physical evidence at Wright Patterson AFB it is possible that it has long since been destroyed; erasing any evidence of Aliens from other galaxies having visited the Earth. An actual physical Alien specimen is needed to confirm their existence. I believe that if the photographs given to me (and published in this book) are genuine, then there is likely remaining physical evidence that can be analyzed genetically to finally put the question to rest.

The release of more documents by the Government would probably only raise more questions but it may pinpoint the location of existing physical evidence; it is only when hard physical evidence is examined and understood that humanity can move forward in the next step towards interstellar community and harmony.

Afterword

This book is published independently with no marketing or advertising assistance or budget. If you can share a link or 'like' on your Facebook Page, Website or Blog; it would greatly help the author in getting this information out into the public eye. It is only with your help that we can move towards an increased knowledge and awareness of our universal community. I hope this book contributes in some small way to the field of Ufology and has been helpful to you the reader.

As this is the first edition, I apologize for any spelling and/or grammatical mistakes that exist and which will be corrected in future editions.

~ Sean Keyhoe

PS: Your comments are welcome at:

seankeyhoe@gmail.com

APPENDIX A

Original use of the term "Flyer Saucer" and events before Roswell

The Roswell incident occurred in early July, 1947 (July 7th) but most people are unaware of the significant event that occurred just two weeks prior. The details of which were documented by Diana Palmer Hoyt in her thesis work on UFOs[7]:

"On June 24, 1947, Kenneth Arnold, a 32-year old successful businessman from Boise, Idaho, was making a routine flight from Chehalis, Washington, to Yakima, Washington, in his private plane, a Callair Aircraft (Steiger 1976, p. 23). Arnold had spent the earlier part of the afternoon installing equipment at the Chehalis, Washington Central Air Service, where he learned of a

---

[7] Thesis submitted to the Faculty of the Virginia Polytechnic Institute and State University in partial fulfillment of the requirements for the degree of Master of Science in Science and Technology Studies.

$5000 reward for the discovery of a C-46 Marine transport that had gone down in the mountains. Arnold decided to make a detour by Mt. Rainier to see if he could find the wreckage. In the course of the search, as he made a 300-degree turn above the town of Mineral, Washington, he was alarmed by a "tremendously bright flash" apparently in front of him.

There was, however, nothing directly in front of him; he looked to his left and to the north, and spotted the source of the flash. A formation of bright objects moved roughly south from Mount Baker toward Mt. Rainier, flying in formation. There were nine shiny objects. He later recalled: "They were flying diagonally in an echelon formation with the larger gap in their echelon between the first four and the last five."

Arnold was at 9200 feet and estimated their speed to be approximately 1700 mph—roughly twice the speed of sound. He later said that he took 500 mph off the estimate because he just couldn't believe the speed. I especially noted that they were all individually independent. They were flying on their own, but every once in a while, they would give off a flash and gain a little more altitude or deviate just a little bit from the echelon formation. This went on all among the nine craft I was observing, alternating periodically, but not in a regular rhythm, I should say

What startled him most was that he could find no evidence of a tail on them. When asked by the clamoring press what they looked like, Arnold said that their motion reminded him of a flat rock as it skipped across water; he later told the press that the objects flew like saucers would if you skipped them across water."

Thus was born the term "flying saucers".

APPENDIX B

NASA Report on UFOs from 1978

NASA RPORT ON UFOs FROM 1978

UNIDENTIFIED FLYING OBJECTS

The information contained here has been compiled to respond to queries on Unidentified Flying Objects directed to the White House as well as NASA.

NASA is the focal point for answering public inquiries to the White House relating to UFOs. NASA is not engaged in a research program involving these phenomena, nor is any other government agency.

BACKGROUND

In July of 1977, Dr. Frank Press, Director of Science and Technology Policy, Executive Office of the President, wrote to Dr. Robert A. Frosch, the NASA Administrator, suggesting NASA should answer all UFO-related mail and als6 to consider whether ~ NASA should conduct an active research program on UFOs. In a letter dated December 21, 1977, Dr. Frosch agreed that NASA will continue to respond to UFO-related mail as it has in the past and, if a new element of hard evidence that UFOs exist is brought to NASA's attention from a credible source, NASA will analyze the unexplained organic or inorganic sample and report its findings.

Quoting from Dr. Frosch's December 21 letter: "If some new element of hard evidence is brought to our attention in the future, it would be entirely-appropriate for a NASA laboratory to analyze and report upon an otherwise unexplained organic or inorganic sample: we stand ready to respond to any bona fide physical evidence from credible sources. We intend to leave the door clearly open for such a possibility."

"We have given considerable thought to the question of what Else the United States might and should do in the area of UFO research. There is an absence of tangible or physical evidence available for thorough laboratory analysis. And, because of the absence of such evidence, we have not been able to devise a sound scientific procedure for investigating these phenomena. To proceed on a research task without a sound disciplinary framework and an exploratory technique in mind would be wasteful and probably unproductive."

"I do not feel that we could mount a research effort without a better starting point than we have been able to identify thus far. I would therefore propose that NASA take no steps to establish research in this area or to convene a symposium on this subject.

"I wish in no way to indicate that NASA has come to any conclusion about these phenomena as such; institutionally, we retain an open mind, a keen sense of scientific curiosity and a willingness to analyze technical problems within our competence."

Reports of unidentified objects entering
United States air space are of interest to
the military as a regular part of defense
surveillance. Beyond that, the U.S. Air
Force no longer investigates reports of UFO
sightings.

This was not always the case. On December
17, 1969, the Secretary of the Air Force
announced the termination of Project Blue
Book, the Air Force program for UFO
investigation started in 1947.

The decision to discontinue UFO
investigations, the USAF said, was based on:
(1) an evaluation of a report (often called
the Condon Report) prepared by the
University of Colorado and entitled
"Scientific Study of Unidentified Flying
Objects;"

(2) a review of the University of Colorado
report by the National Academy of Sciences;
(3) past UFO studies; and (4) Air Force
experience investigating VFO reports for two
decades.

As a result of these investigations and studies, and experience gained from investigating UFO reports since 1948, the conclusions of the Air Force were: (1) no UFO reported, investigated, and evaluated by the Air Force has ever given any indication of threat to our national security; (2) there has been no evidence submitted to or discovered by the Air Force that sightings categorized as "unidentified" represent technological developments or principles beyond the range of present day scientific knowledge; and (3) there has been no evidence indicating that sightings categorized as "unidentified" are extraterrestrial vehicles.

With the termination of Project Blue Book, the Air Force regulation establishing and controlling the program for investigating and analyzing UFOs was rescinded. All documentation regarding the former Blue Book investigation has been permanently transferred to the

Modern Military Branch, National Archives and Records Service, 8th Street and Pennsylvania Avenue, N.W., Washington, DC 20408, and is available for public review and 'analysis. Those wishing to review this material may obtain a researcher's permit from the National Archives and Record Service.

Also available: Scientific Studv of Unidentified Flving Objects. Condon report study conducted by the University of Colorado under contract F44620-76-C-0035. Three volumes, 1,465p. 68 plates.

Photoduplicated hard copies of the official report may be ordered for $6 per volume $18 the set of three, as AD 680:975, AD 680:976, and AD 680:977, from the National Technical Information Service,
U.S. Department of Commerce, Springfield, VA 22151.

Review of University of Colorado Report on Unidentified Flying Objects.

Review of report by a panel of the National Academy of Sciences. National Academy of Sciences, 1969, 6P Photo-duplicated hard copies may be ordered for $3 as AD 688:541 from the National Technical Information Service, U.S. Department of Commerce, Springfield, VA 22151.

NASA is aware of the many UFO reports made in recent years.-However, the majority of inquiries to NASA concerning UFO sightings address themselves to the reported sightings by astronauts during Earth orbital and lunar missions and the report by President Carter while serving as Governor of Georgia.

During several space missions NASA astronauts reported phenomena not immediately explainable. However, in every instance NASA satisfied itself that what had been observed was nothing which could be termed abnormal in the space environment. The air-to-ground tapes of all manned missions are available at the Johnson Space Center, Houston, for review by the serious researcher.

On October 12, 1973, while serving as Governor of Georgia, Mr. Carter responded to inquiries from the National Investigations Committee on Aerial Phenomena (NICAP) saying that he had seen a bright, moving object in the sky over Leary, Georgia, in October of 1969. He said the object was visible for 10 to 12 minutes and, at one point, shone as brightly as the Moon.

(Author's note: I thought this was extremely interesting to see a UFO report from former President Jimmy Carter that has not received much attention.)

The regional NICAP representative investigated the sighting and reported there was no evidence to support anything beyond placing what Mr. Carter saw in NICAP's "unidentified" category. However, it has

been suggested by some students of aerial
phenomena that Mr. Carter may have viewed
the Planet Venus which, at certain times,
may appear many times brighter than a star
of the first magnitude.

Since NASA *is* not engaged in day-to-day UFO
research, it does not review UFO-related
articles intended for publication, evaluate
UFO-type spacecraft drawings or accept
accounts of UFO sightings or applications
for employment in the field of aerial
phenomena investigation. All such material
will be returned with NASA's thanks to the
sender.

A number of universities and scientific
organizations have considered UFO phenomena
during periodic meetings and seminars. In
addition, a number of private domestic and
foreign groups continue to review UFO
sighting reports actively. Some of these
organizations are:

(1) National Investigations Committee on
Aerial Phenomena John L. Acuff, Director

Suite 23 3535 University Boulevard, West
Kensington, MD 20795 (301) 949-1267

(2) The Committee for the Scientific
Investigation of Claims of the Paranormal
UFO Subcommittee Robert Sheaffer, Chairman
9805 McMillan Avenue Silver Spring, MD 20910
(301) 589-8371

(3) Aerial Phenomena Research Organization
James and Coral Lorenzen, Directors 3910 E.
Kleindale Road Tucson, AZ 85712
(E02) 793-1825

(4) Mutual UFO Network Walter H. Andrus Jr.,
Director 103 Old Towne Road Seguin, TX 78155
(512) 379-9216 (5) The Center for UFO Studies
Dr. J. Allen Hynek, Director 924 Chicago
Avenue Evanston, IL 60202 (312) 491-1780

APPENDIX C

MUFON CONCLUSIONS ON STEPHENVILLE UFO EVENT

There are several conclusions that the authors have reached with this report and its supporting analysis. The first and primary conclusion is that there was definitely a real and physical object that appeared and was witnessed on January 8, 2008, in the Dublin-Stephenville area. Reports of unidentified flying objects occur all the time. Most of those reports are from single individuals or a group of individuals who see an unexplained object at a given location and time. These types of reports are easier to explain away because there is usually, whether likely or not, some type of explanation that can be constructed to explain away the event at a specific time and place.

What makes the Dublin-Stephenville event unique is that there are multiple witnesses at different locations and the sightings occur over a three hour time period.

Additionally, radar data identifies unknown aircraft in the sky at the same time as many of the witness sightings. So in the Dublin-Stephenville case, one would have to attempt several varied low probability explanations to try and explain away all of the various sightings. The likelihood that all of these witnesses miss-identified separate known objects at different times, in different but closely associated geographic locations, all within a 3 ½hour time period is extremely low. It is much more reasonable to believe that these witnesses truly saw an object that could not be explained by any objects with which they are familiar.

As to what these witnesses saw, it is difficult to determine. It was not any known aircraft. The enormous size of the object, its complete silence, and its ability to travel at high rates of speed and to also remain stationary or travel at slow speeds, is not explained by any known aircraft. The smallest size calculated from witness descriptions was 524 feet and most of the calculations based on approximate distance of the object and witness descriptions of degrees of sky covered by the object indicated an object closer to 1,000 feet in size. Twice, radar picked up an unknown object flying at 1,900-2,100 mph.

Admittedly, it could have been a coincidental radar hit...but in both cases that coincidence occurred when a witness saw a very fast moving object in the same direction as an object painted by radar. Twice, radar tracked slow moving objects, for extended periods of time, that were very near the witness' location, in the direction described by the witness, and at approximately the same time that the witness saw the unknown object. It is very difficult to dismiss witness testimony that is corroborated by radar. And to further augment the strangeness of these events, radar tracked one of those two objects for over an hour as it traveled directly toward Crawford Ranch.

The authors cannot comment on the source or origin of this object, but it is clear to the authors that the unknown object was real and not imaginary.

The second conclusion of this report is that the military did not react overtly to the presence of these unknown objects. In light of the disaster of Sept. 11, 2001, the authors of this report have concerns with how the military reacted to an unknown aircraft(s) in U.S. air space. It is clear that there was an unknown object without any transponder beacon traveling along a path that began south of Dublin and that proceeded on a direct path to Crawford Ranch. This object was tracked by the FTW radar for over an hour. Military jets flew within a mile of this object on their way to the Brownwood MOA. The F-16s had to have seen this object on their radar and the suspected AWACS that was circling this area must have detected and recorded this object on its state-of-the-art radar. This must have raised concerns, yet the radar tracks of the military jets, indicates that there was no reaction by them to this object during the hour of time in question. What could explain this lack of reaction?

One possibility is that the military knew the identity of the object and instructed the F-16 pilots to ignore it and stay on course to the MOA. But this possibility is countered by all of the military replies to the FOIAs that indicated the military had no aircraft in the area other than the F16s from CAFB that have already been identified. Secondly, if it was a military aircraft then it was violating FAA and military MARSA rules by not having a transponder beacon code activated while being outside of a MOA. This leaves us with the possibility that the military either did not see the object or just ignored it. In light of what happened on 9/11, what if the unknown object had been a terrorist aircraft?

The Air Force should explain what their radar detected on the evening of January 8, 2008, and the reason as to why the military jets in the area did not react. The third conclusion is that military aircraft traffic in the area was unusually heavy and twice military aircraft strayed out of their standard Military Training

Routes and into civilian airspace. Ten F-16 jets from Carswell AFB were documented as flying into the Dublin-Stephenville area within a 2 hour time period as well as a probable AWACS that circled the area for over 4 hours. A FOIA requesting information to determine how unique this level of jet activity may be was sent to the 10thAir Force in Ft. Worth on May 7th, 2008. An acknowledgement of the correspondence has been received but a formal reply is still pending. Two CAFB sorties, a total of 4 F-16s, returning to CAFB belatedly activated military beacons and veered unexpectedly eastward over Stephenville toward DFW civilian aircraft arrival traffic patterns.

There is no explanation as to why the military jets strayed from their normal MTR. Since they did not initially leave CAFB with beacons, it is reasonable to assume that something occurred that caused those aircraft to break away from their lead aircraft and request a beacon code so that they could veer away from the standard MTR. The last conclusion is that there are indications that requests submitted under the Freedom of Information Act are not considered seriously by the U.S. military and were completely ignored by the Dept of Homeland Security's branch, U.S. Customs & Border Patrol. If true, this would be a violation of a law passed by the Congress of the United States. FOIA requests are usually handled by a clerk who is an intermediary between the submitter and who ever within the military decides what information can be provided. The reply is uniformly the same from military base to military base. The standard reply has obviously been crafted specifically in the manner that the military should use to deal with FOIAs from the public. The standard reply is, "There are no responsive records that meet your request".

With the events of September 11, 2001, it is understandable that the military would choose caution in dealing with any release of information regarding their operational activities. But in this case, we are discussing military activities within the United States, during a four hour period of time, on a specific date, and in a small and specific area of Texas. One would be hard pressed to argue that release of this type of information would be a threat to national security. And exactly what complicated information was requested? Only the following..."Do you have any evidence to support if Military Base "X" was flying aircraft within 50 miles of Stephenville, Texas, on Jan.8, 2008?" "Can you provide copies of radar images from any military aircraft operating with 50 miles of Stephenville on Jan 8, 2008?" Surely the military can say, "No, we had no aircraft in the area.", or perhaps, "We cannot release this information due to reasons of national security." But, no, instead we receive..."We found no documents responsive to your request."

On the other hand, we would like to again express our sincere thanks to the National Weather Service and the Federal Aviation Administration for their excellent responses and their willingness to abide by the requirements of the Freedom of Information Act. They communicated effectively and if they did not have the required information, they readily said so.

We are a nation of freedom that is based on a set of principals designed to maintain our individual liberties. When our government bodies reach a point that they do not feel compelled to honor the requests of their citizenry, as defined by the laws of this nation, we have taken a path that allows the government to arbitrarily and secretly decide what we should and should not know. **The American people have a right to know what did or did not occur on January 8, 2008 in the Dublin-Stephenville area.**

Share this information...God speed

www.ingramcontent.com/pod-product-compliance
Lightning Source LLC
Chambersburg PA
CBHW051514170526
45165CB00002B/473